# Nylon Plastics Technology

# PLASTICS AND RUBBER INSTITUTE MONOGRAPHS

The Plastics and Rubber Institute through its Books Sub-committee is responsible for selecting the title and scope of the subject matter of a monograph, in consultation with the author. Each manuscript is appraised by specialist readers and prepared for publication by an editor selected by the Institute. The Sub-committee gives initial guidance to the author and, through its readers, checks and revises the manuscript. Thus each monograph gives a critical, balanced, up-to-date treatment of the subject that is as far as can be ascertained factually correct and, above all, authoritative. Dr. C. A. Redfarn is the editor of this monograph.

# Nylon Plastics Technology

W. E. Nelson, BSc, PhD, CChem, FRIC

Published for
The Plastics and Rubber Institute

**NEWNES—BUTTERWORTHS**
LONDON   BOSTON
Sydney  Wellington  Durban  Toronto

THE BUTTERWORTH GROUP

UNITED KINGDOM
Butterworth & Co (Publishers) Ltd
London: 88 Kingsway, WC2B 6AB

AUSTRALIA
Butterworths Pty Ltd
Sydney: 586 Pacific Highway, Chatswood, NSW 2067
Also at Melbourne, Brisbane, Adelaide and Perth

CANADA
Butterworth & Co (Canada) Ltd
Toronto: 2265 Midland Avenue, Scarborough, Ontario M1P 4S1

NEW ZEALAND
Butterworths of New Zealand Ltd
Wellington: 26–28 Waring Taylor Street, 1

SOUTH AFRICA
Butterworth & Co (South Africa) (Pty) Ltd
Durban: 152–154 Gale Street

USA
Butterworth (Publishers) Inc
Boston: 19 Cummings Park, Woburn, Mass. 01801

First published 1976

Published for the Plastics and Rubber Institute
11 Hobart Place, London SW1W 0HL

© The Plastics and Rubber Institute, 1976

All rights reserved. No part of this publication may be
reproduced or transmitted in any form or by any means,
including photocopying and recording, without the written
permission of the copyright holder, application for which
should be addressed to the publisher. Such written
permission must also be obtained before any part of this
publication is stored in a retrieval system of any nature.

This book is sold subject to the Standard Conditions of
Sale of Net Books and may not be resold in the UK below the
net price given by the publishers in their current price list.

ISBN 0 408 00251 4

*Filmset and printed by Thomson Litho Ltd, East Kilbride, Scotland*

# Preface

The nylons emerged as commercial engineering thermoplastics after the Second World War following their remarkable wartime success in the textile field. The large-scale production of nylon plastics grew mainly from the adaption of conversion methods and equipment already used for other thermoplastics, together with the availability of relatively cheap raw materials for bulk manufacture of the polymers. The premium properties of the nylons quickly led to their acceptance and use as high-grade engineering plastic materials.

In this volume an attempt has been made to present, for the use of students taking technology courses in colleges and for technologists in the nylon industry, a broad but balanced picture of present-day nylon plastics technology. The chapters follow the sequence adopted for other monographs on plastic materials in this series, and cover raw materials, chemistry, properties, conversion processes and applications, with an historical introduction and a final chapter on analysis and testing.

While the emphasis is largely on the technology of the nylons, there is coverage of the scientific aspects and theory sufficient for the needs of most technologists. In discussing properties the need is emphasised to obtain and use reliable design data to exploit fully the unique combination of properties possessed by nylons.

In compiling this book a large number of references were consulted, and those from which data were selected for inclusion are listed at the end of each chapter. Particular acknowledgement must be made to Carl Hanser Verlag, publishers of *Kunststoff-Handbuch*, for permission to reproduce a number of diagrams and tables from Vol. VI (*Polyamide*); also to Imperial Chemical Industries Ltd, Plastics Division, E. I. Du Pont de Nemours and Co. (Inc), Plastics Department, the Polymer Corporation, and Polypenco Ltd, for permission to make extensive use of information and diagrams appearing in their technical service notes, brochures and manuals.

Of the many individuals who assisted during the preparation of the monograph the author would like to thank particularly Mr P. C.

Powell of the Institute of Polymer Technology, Loughborough, who checked the manuscript and provided useful advice and criticism on Chapter 4; Dr J. Maxwell, of Imperial Chemical Industries Ltd, Plastics Division, who did likewise for Chapters 5 and 6; Dr C. Goebel, formerly of Research Department, Polymer Corporation, for checking Chapter 3; Mr D. Lawson, formerly of *European Chemical News*, who provided much of the data for Chapter 2; and finally Dr C. A. Redfarn, the Plastics and Rubber Institute editor of this series, for his expert guidance throughout the preparation of the monograph. Acknowledgement is made of the support given by the Directors of Polypenco Ltd at the manuscript stage, and especial thanks are due to Mrs Hecks and Mrs Barnes of Polypenco Ltd, who carried out most of the typing.

W.E.N.

# Contents

**1 Introduction   1**

1.1 General
1.2 Historical
1.3 Present status of the industry
1.4 Future developments
References

**2 Raw materials for polyamides   10**

2.1 Introduction and terminology
2.2 Economics
2.3 Feedstocks and intermediates
References

**3 Basic chemistry and manufacture of polyamides   26**

3.1 Classification and nomenclature
3.2 Direct stepwise polymerisation
3.3 Ring scission polymerisation—hydrolytic and anionic polymerisation
3.4 Other stepwise polymerisation mechanisms
3.5 Other commercial nylons—mechanisms and technical manufacture
3.6 Polyamides from aromatic diamines
3.7 Copolyamides
3.8 Cyclic oligomers
3.9 Preparative methods, based on physico-chemical principles used
3.10 Molecular weight control
References

**4 Properties of polyamides   55**

4.1 Chemical properties
4.2 Mechanical properties of dry polyamides
4.3 Moisture absorption and its effect on mechanical properties
4.4 Physical properties
References

## 5  Conversion processes    145

5.1  General
5.2  Injection moulding
5.3  Extrusion
5.4  Melt spinning
5.5  Casting
5.6  Powder processing
5.7  Machining and finishing operations
    References

## 6  Applications of polyamides    188

6.1  General
6.2  Mechanical applications
6.3  Electrical applications
6.4  Chemical applications
6.5  Miscellaneous applications
6.6  Choice of production method
6.7  Applications of modified polyamides
    References

## 7  Characterisation, analysis and testing of polyamides    205

7.1  Features characterised
7.2  Methods of characterisation
7.3  Identification and analysis
7.4  Testing for quality, type and design
    References

## Index    227

# 1
# Introduction

## 1.1 General

The ever-increasing use of plastics in civilised society is now recognised as a significant contributory factor to an increased standard of living. Indeed it could be said that a country's prosperity can be measured by its plastics production in much the same way as sulphuric acid production was quoted as the index in the early years of this century.

The establishment of large-scale plants for manufacture of plastics materials (particularly in the United States but also in Europe, Japan and elsewhere) made available for world markets in the 1960s an abundance of cheap but high-quality material that could be converted to the shapes, forms and finished products familiar to all of us today. Generally the technology that led to this growth is now well known, and international and national companies are producing more or less standard plastic materials using techniques either developed by themselves or obtained by licence agreements from others.

The economic status of polyamides in relation to other plastics can be assessed from published statistics of sales, production and usage; these are presented in detail in section 1.3, and it is clear from them that polyamides account for only a small percentage of total plastics production.

The unique position of polyamides in the plastics field is, however, due largely to their high strength and toughness, which properties bring them into the sphere of conventional engineering materials. Allied to this is their high melting point yet thermoplastic behaviour, which allows of their processing by conventional extrusion and moulding.

The aim of this monograph is to review the science and technology of polyamides with special emphasis on their properties and applications. An account of the raw materials used in the manufacture of polyamides is included to show their relationship to organic chemicals generally.

## 1.2 Historical

The decision of E. I. Du Pont de Nemours & Co. (Inc.) to embark on fundamental chemical research in 1928 marked the beginning of a new age of chemical discovery for the company[1]. To initiate this work Du Pont appointed Dr Wallace N. Carothers of the University of Illinois. Carothers, who at 32 had a high reputation as an organic chemist, was allowed to select his own line of research and was given ample funds to carry it out. Carothers was fascinated by the natural polymers in the animal and vegetable kingdoms—for example, wool and silk. His proposal was to try to synthesise them. Using dibasic organic acid and di-hydroxy alcohol, Carothers was able to synthesise large molecules showing colloidal behaviour, but real success came in April 1930: his colleague Dr J. W. Hill, by using the technique of the molecular still to remove effectively the water produced by condensation, was able to produce a tough elastic material which, although it could be drawn into a fibre, was still brittle. Carothers named these materials 'super-polymers', but because of their low melting point and their solubility in common organic solvents he deemed them unworthy of further detailed investigation. So, switching his attention to polyester amides and polyamides, he found that these materials conformed more to his ideal 'super-polymers'. In May 1934 he demonstrated the production from polyamides of fibres that were heat resistant and stood up to washing and dry cleaning. Many more polyamide 'super-polymers' followed and by 1935 polyamide 66, better known as nylon 66, was judged to be the material with most promise. The pure research on the super polymers was succeeded by applied research on polyamides, particularly nylon 66. A pilot-plant was built to produce monofilaments and film. On 27 October 1938 Du Pont announced their development of a new 'group of synthetic super-polymers' from which could be produced high-strength textiles. The super polymers were said to be protein-like in character and the generic name 'nylon' was applied to them.

The properties of the new material caught the imagination of the American public when it was demonstrated that the fibre had superior mechanical properties to natural silk. Ladies' stockings made from the new material were exhibited at the New York World Fair in 1939 and in a Chicago Fair in the same year. In May 1940 nylon stockings were on sale in stores in the US. Demand soon exceeded supply, and plants to produce the nylon polymer and yarn were erected in many parts of the US. At this time the market was

## Introduction

mainly for hose, toothbrush bristles and sports gear, but many other textile applications were found.

The outbreak of hostilities with Japan in 1941 cut off the importation into the US of Japanese silk, but fortunately nylon was there to take its place.

Carothers died in April 1937 at the age of 41, so did not live to see the full realisation of his polymer research work. During his nine years with Du Pont more than fifty patents were issued in his name and about as many scientific papers[2].

While Carothers was working in the States, laboratory work on polycondensation of esters and amides was being carried out by I. G. Farben in Germany. As with Carothers no useful fibre-forming materials were found in the early work, and during the economic depression of the 1930s work was suspended. The disclosures by Du Pont, in patents and other publications, of success in the field of fibre-forming materials led to resumption of the German work. Success came when Schlect in 1938 showed that by careful selection of catalysts—including aminocarboxylic acids, 66 salt and water—the caprolactam ring could be opened and straight-chain polymerisation effected. The stability and ease of purification of caprolactam brought success in the early experiments and high polymers were produced with properties very similar to nylon 66. The high polymers from caprolactam were designated 'Perlon'.

In 1939 a licence agreement was arranged between I. G. Farben and Du Pont, who meantime had published a large number of patents on the spinning and drawing of polyamides[3]. With the possession of much of Du Pont's know-how in fibres the pace of development quickened, and in 1939 the first production form of 6-polyamide (termed 'Perluranborsten') was produced in I. G. Farben's Berlin-Lichtenberg factory and became commercially available. Technical work continued during the Second World War, and injection-moulded parts in nylon 6 and 66 were produced and tested in service. The work at I.G.'s Ludwigshaven plant was on developing the use of nylon as a plastic material rather than as a fibre, which had hitherto been Du Pont's main nylon product.

After the war Du Pont quickly adapted their market to the plastic as opposed to the fibre uses of nylon. Other countries—the UK, France and Holland—quickly followed the two leading nations in polyamides and set up plants for both fibre and plastics production of polyamides.

The above historical outline of the discovery and development of polyamides would be incomplete without mention of Carothers' basic ideas on molecular chain growth in polymerisation[4]. He believed that the synthesis of linear giant molecules could only be achieved by head-to-tail condensation reactions involving small molecules each of which had bifunctionality, i.e. one reactive group at each end of the molecule. Carothers defined functionality in a monomer as an arrangement of atoms that could lead to a reaction step, e.g.—OH, —NH$_2$, —COOH and so on. According to the number of such units in the monomer (one, two or more) the monomer is called monofunctional, bifunctional, etc. Carothers discovered that, if such bifunctional monomers were used that on reaction could form five- or six-membered rings by intramolecular reaction, this was the preferred mode of action. However, if rings larger than six were possible by intramolecular reaction, linear reaction occurred preferentially. The Sachse-Mohr theory may be invoked in support of the observation of the preferential formation of five- or six-membered rings from bifunctional monomers.

The use of the molecular still was undoubtedly the basis of Carothers' success in producing useful high-molecular-weight polymers. Previously molecular weights of only about 5000 had been achievable. Using the still the molecular weight could be raised to over 10 000 because the low rate of diffusion of volatile products of condensation and the tendency for them to remain dissolved or absorbed could be overcome.

Carothers had to choose for detailed study either the tougher, less easily fusible and less soluble of the polyamides he obtained from ε-aminocaproic acid or the strong, pliable, highly-oriented (but low-melting-point and soluble) polyester-polyamides. His concentration on the mixed super-polymers may have been the reason for his failing to develop polycaprolactam as a useful fibre, thus allowing the Germans time to develop the catalytic system which was used to polymerise ε-aminocaproic acid so successfully in the work which eventually produced 'Perlon'.

## 1.3 Present status of the industry

The growth of the plastics industry, which includes polyamides, is conveniently observed in published statistics. Those believed to be accurate and reliable are published annually in *Europlastics* and *Modern Plastics*; figures are available on sales, production and consumption of all nylons, allowing year-by-year comparison to show growth rate. The data refer only to the US and the UK, the former being by far the world's largest producer of nylon plastics. The trends in growth-rate and end-use of nylons throughout the world can be expected to follow closely the American and British pattern.

Table 1.1 shows the total sales of nylon in the US in the period 1962–1973. (In this and other tables in this section all quoted figures have been converted to units of 1000 tonnes to aid comparison.) Thus for the US alone, the major producing country, sales more than quadrupled during that time.

**Table 1.1** SALES OF NYLON (ALL TYPES) IN US, 1962–1973 (data from *Modern Plastics*, annual January issues)

| Year | 1962 | 1963 | 1964 | 1965 | 1966 | 1967 | 1968 | 1969 | 1970 | 1971 | 1972 | 1973 |
|---|---|---|---|---|---|---|---|---|---|---|---|---|
| Total sales (1000 t) | 18.6 | 20.1 | 22.3 | 27.2 | 27.6 | 27.7 | 34.5 | 40.2 | 46.0 | 49.9 | | |
| | | | | | | | *New basis for reporting* | | | 58 | 67 | 80 |

**Table 1.2** SALES OF PLASTICS IN US IN 1972 AND 1973 ALLOCATED TO TYPE (data from *Europlastics* (British Plastics Edition), Jan. 1973, 1974; *Modern Plastics*, Jan. 1973, 1974)

| Thermoplastics type | Sales (1000 tonnes) | |
|---|---|---|
| | 1972 | 1973 |
| Polyolefines | 4200 | 4896 |
| PVC (Polymers) | 1975 | 2171 |
| Polystyrene | 2111 | 2407 |
| Polyvinylacetate | 370 | 391 |
| Cellulosics | 75 | 77 |
| Acrylics | 208 | 233 |
| *Thermosetting type* | | |
| Aminos | 411 | 464 |
| Alkyds | 290 | 342 |
| Phenolics | 651 | 654 |
| Epoxides | 83 | 99 |
| Polyesters | 416 | 496 |
| Polyurethanes | 493 | 593 |
| Miscellaneous* | 110 | 122 |
| Total | 11393 | 12945 |

*Includes acetal, fluoroplastics, nylon, polycarbonates, silicones, etc.

There are no reliable statistics that show the proportion of nylon sales to total plastics sales. Some indication is, however, given in *Table 1.2* of sales in the US for 1972 and 1973 for all plastics. The miscellaneous group, which includes nylon, comprises only 9.7 per cent of total sales in 1972 and 9.5 per cent in 1973. It is evident that on sales tonnage in the US nylon ranks very low among plastics materials. Expressing sales on a cost basis would considerably improve the ranking of nylon since it is a relatively high cost material. However, polyamides as a class are and will remain for a long time low-volume, high-cost plastics.

The allocation of nylons to particular end-uses, shown for UK and US consumption in *Table 1.3*, indicates one reason for the comparatively low tonnage of nylon sales and production compared with other plastics. The table shows the major uses to be for automobiles

**Table 1.3** NYLON CONSUMPTION IN UK AND US IN 1972, 1973: ALLOCATION TO END-USE (data from *Europlastics* (British Plastics Edition), Jan. 1973, 1974; *Modern Plastics*, Jan. 1973, 1974)

|  | Consumption (1000 tonnes) | | | |
|---|---|---|---|---|
|  | UK | | US | |
|  | 1972 | 1973* | 1972 | 1973 |
| *End-use* | | | | |
| Appliances | | | 5.3 | 6.1 |
| Automobiles and trucks | | | 13.9 | 16.0 |
| Building and construction | 0.2 | | | |
| Land transport | 1.4 | | | |
| Consumer products | 1.5 | | 3.9 | 5.0 |
| Electrical and electronics | | | | |
| (including communications) | 0.4 | | 9.0 | 11.5 |
| Fancy goods | 0.1 | | | |
| General industry | 3.1 | | | |
| Machinery parts | | | 6.6 | 9.2 |
| Housewares | 2.1 | | | |
| Others (including exports) | | | 8.0 | 10.6 |
| *Extrusions* | | | | |
| Filaments | | | 6.9 | 7.2 |
| Film | | | 5.7 | 5.0 |
| Sheet, rod and tube | | | 3.4 | 4.0 |
| Wire and cable | | | 3.8 | 4.2 |
| Totals | 8.8 | | 66.5 | 78.8 |

*Statistics not available

and trucks, electrical and general industrial outlets. These are, of course, generally engineering applications. The adoption of plastics as metal replacements in established all-metal designs has been slow mainly because of lack of engineering data and practical experience. Confidence is rapidly growing in this field as design data becomes more readily available, and the use of polyamides in automobiles is rapidly increasing. The development of reinforced grades of nylon and new fabrication techniques will undoubtedly accelerate progress.

The use of nylons in housewares must not be overlooked, and in applications where its lightness, toughness, strength and temperature resistance can be used with advantage it will be accepted.

The figures given in *Table 1.4* show the all-nylon consumption in the UK and US during the years 1972 and 1973. The per capita consumption based on the population of the two countries (UK population in 1971 census 55 million, US population estimated in 1970 at 203 million) shows an almost equal usage in the two countries.

**Table 1.4** NYLON CONSUMPTION IN UK AND US IN 1972, 1973 (data from *Europlastics* (British Plastics Edition), Jan. 1973, 1974; *Modern Plastics*, Jan. 1973, 1974)

|  | 1972 | | 1973 | |
| --- | --- | --- | --- | --- |
|  | UK | US | UK | US |
| Total consumption (1000 t) of all nylon | 18 | 66.5 | 21 | 78.9 |
| Consumption (1000 t) per $10^6$ capita of population in each country | 0.33 | 0.33 | 0.38 | 0.39 |

The observations from the statistics tabulated are the author's own and without more detailed knowledge of their derivation it would be invalid to make more than the broadest generalisations.

## 1.4 Future developments

Because of the rapid pace of innovation in plastics it would be unwise to predict developments in polyamides beyond the mid-1980s. For the decade to 1985 likely developments in raw materials, processes, and fabrication or conversion techniques are briefly discussed in the following paragraphs.

## 1.4.1 MATERIAL DEVELOPMENT

There is unlikely to be a major improvement in the present range of properties of polyamides through the synthesis of new members of the class, because all the obviously feasible arrangements around the carbo-imide group characteristic of the class seem to have been examined already.

A great deal of effort, however, will be put into improving the present properties, and a much wider range of nylon polymers, each tailored to a specific application, should become available. These will include heat-resistant types containing aromatic groupings along the chain, fire-resistant types, and types of higher deflection temperature—produced for example by controlled cross-linking using appropriate catalysts, or by external means such as high energy radiation.

The present trend of modifying polyamides by the use of fillers will continue, but advances in this area will depend upon success in solving the problems associated with filler-matrix compatibility, and filler adhesion and dispersion. In the engineering field the design parameters of any new material must be established as early as possible, preferably before marketing, so that designers and engineers can adopt the material with a reasonable degree of confidence.

Environmentally resistant types of polyamides will be generally available when the present generation of UV heat and hydrolysis-stabilised types that have been patented in the last few years have been critically examined under service conditions, and those that have stood the test of time have been selected for further development.

## 1.4.2 RAW MATERIAL PROCESS DEVELOPMENT

Novel processes for producing the present or new polyamide types will continue under active investigation. The preoccupation with size for the fabrication of large engineering parts puts a strain on the traditional methods of fabrication through injection moulding and extrusion, and the direct conversion method examplified in the process of anionic polymerisation of caprolactam is regarded with great favour. Similar systems stand a good chance of being commercially successful as the result of stimulating competition from other similarly derived polymers, such as the solid polyurethanes.

## 1.4.3 CONVERSION AND FABRICATION DEVELOPMENT

With the hardening in design in injection moulding and extrusion, machinery development in this area will be concerned with increased size and sophistication. Major advances are unlikely. The ICI sandwich moulding process[5], which injects two polymer formulations into the mould sequentially to form a composite structure of a core and skin each with its own properties, is likely to be fully developed for polyamides. The Engel process[6], which uses ultra-high pressure and catalysts to cross-link linear polymer chains, might well be applied to polyamides if suitable chemical techniques are found.

Rotational moulding will be more widely used, particularly for larger fabrications.

The techniques of solid-phase forming of stock shapes and fabricated parts, which are slowly improving as the basic problems are being solved, should permit further useful alternatives in the fabrication methods available for polyamides.

For small precision parts requiring good dimensional stability, the technique of powder sintering (developed originally for metallic components) will be used more generally for polyamides in the replacement of metals by plastics for automotive parts and for small mechanical components.

### REFERENCES

1 Dutton, W. S., *Du Pont: one hundred and forty years*, Charles Scribner's Sons, New York (1942)
2 Mark, H. and Whitby, G. S., *Collected papers of Wallace H. Carothers on polymerisation*, Vol. 1, *High polymers*, Interscience (1940)
3 Du Pont de Nemours, US Patents 2 071 250 (16.2.1937), 2 130 523 (20.9.1938), 2 130 947 (20.9.1938), 2 130 948 (20.9.1938), 2 137 235 (22.11.1938), 2 149 273 (7.3.1939), 2 157 116 (9.5.1939), 2 158 064 (16.5.1939), 2 163 584 (27.6.1939), 2 174 619 (3.10.1939)
4 Carothers, W. H. and Hill, J. W., *J. Am. chem. Soc.*, **54**, 1559–1587 (1932)
5 *Plastics and Polymers*, **39**, 288 (1971)
6 *Plastics and Polymers*, **38**, 174 (1970)

# 2
# Raw materials for polyamides

## 2.1 Introduction and terminology

The term nylon is applied to the synthetic polyamides that have fibre, film, and/or plastic-forming properties. All nylons have in the polymer materials the amide (—CO—NH—) group, the link that joins the repeating hydrocarbon units of various lengths. The intermediates for nylons are dicarboxylic acids, diamines, amino acids and lactams. The nylons are usually identified numerically using the number of carbon atoms in the basic units of the polymer chains. A duplex digit is used where the nylon is derived from a diamine and diacid, the first digit of the duplex referring to the diamine. For example, the polymer from adipic acid and hexamethylene diamine is termed nylon 66, that from sebacic acid and hexamethylene diamine nylon 6.10, that from caprolactam nylon 6. The terms nylon and polyamide are used interchangeably in commercial and technological publications.

## 2.2 Economics

The world volume of production of polyamides was until recently closely affected by the price and availability of feedstock for the intermediates; in the latter half of the 1960s, however, production became much more dependent on the ability to sell the product in the form of fibres in competition with the acrylic and polyester fibres, which were gaining a much larger share of the fibres market. At the same time the processes of conversion of raw materials to intermediates (which started with the original commercial processes used by Du Pont and I. G. Farben) continue to be a fertile field of innovation aimed at reducing costs and offering alternative routes to the product.

For new intermediate plants starting up, the choice of route depends both on feedstock price and availability and on the market for by-products—particularly, in the case of polyamides, ammonium sulphate, which for most processes is the by-product produced in greatest amount. Ammonium sulphate is used in the fertiliser industry and the demand for it is subject to long-term variation.

The main commercially-used routes to the nylon intermediates, with their advantages and disadvantages, are described in the following sections.

## 2.3 Feedstocks and intermediates

Feedstocks for the commercial polyamide intermediates are with few exceptions aromatics, cyclohydrocarbons, or olefines derived largely from the petrochemical industry. One exception is castor oil, which is the precursor of both sebacic acid (a component of nylon 6.10) and also undecanoic acid (the intermediate for nylon 11).

### 2.3.1 INTERMEDIATES FOR NYLON 66

#### 2.3.1.1 Adipic acid

The usual process for adipic acid intended specifically for nylon 66 uses the nitric acid oxidation of cyclohexanol or KA oil (mixture of cyclohexanol and cyclohexanone). The latter is reacted with dilute nitric acid (about 60 per cent) at 60–110°C in the presence of a mixed ammonium vanadate catalyst. The yield is claimed to be over 90 per cent and the conversion rate is high.

The advantage of this process is its simplicity and its ability to produce material of high purity at high conversion rate. The disadvantage is the large nitric acid consumption and requirements for nitric acid recovery.

This process can be combined with the process for making KA oil from cyclohexane discussed in greater detail in section 2.3.2.1(a)(ii).

Adipic acid produced by this process can be made the starting material for hexamethylene diamine, the other monomer for nylon 66, although many other processes exist (see section 2.3.1.2).

Considerable development work on other adipic acid processes is described in the technical literature[1].

## 2.3.1.2 Hexamethylene diamine (HMD)

The adipic acid process and the butadiene process described below account for most of the world production of HMD. These and other processes used industrially are shown with flow diagrams in *Figure 2.1*[1]. The considerable number of processes available for HMD

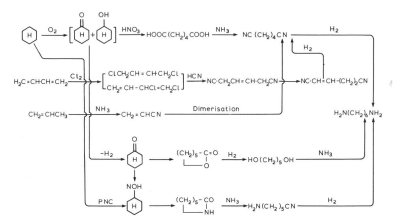

Figure 2.1 Commercial hexamethylene diamine processes (courtesy *Chemical Economy and Engineering Review*)

production indicates the key role of this intermediate in effecting economies in nylon production.

### (a) Adipic acid process

This process has the advantage that the HMD can be produced in a series of steps starting with adipic acid, the other monomer for AH salt (nylon 66 salt). The first step consists of vaporising the adipic acid in ammonia gas and dehydrating the mixed gas in a reactor over a suitable catalyst to obtain the nitrile. Some of the acid is lost by decomposition at this stage and yield is reduced. (Considerable development work protected by patents and aimed at improving the efficiency of this part of the process is being undertaken by several large raw-material producers.) The second step, hydrogenation of adiponitrile, is usually carried out in the liquid phase at high pressure over a nickel or cobalt catalyst. Here again several variations in the

## Raw materials for polyamides

method, involving reactor type, reaction conditions and catalyst, are known.

The adipic-acid route to HMD has no major disadvantages, and the process is used by the majority of the large raw-material suppliers.

### (b) Butadiene processes

The original process was developed and is used commercially by Du Pont. The flow diagram is shown in *Figure 2.1*. Adiponitrile is obtained almost quantitatively and its hydrogenation follows the same steps as in the adipic-acid process. While the butadiene process has the advantage of low cost of butadiene feedstock and good conversions and yield, considerable quantities of hydrocyanic gas and chlorine are used and capital costs are high. To obviate the use of chlorine in this process, Du Pont have recently developed and commercialised a new route from butadiene[2,9] which, it is claimed, results in a saving of 15 per cent over their older process for making adiponitrile.

The reaction steps in the improved process are shown below in diagrammatic form:

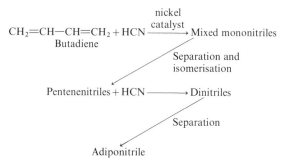

Another route to adiponitrile using butadiene as starting material has been developed by Esso Research and Engineering[2]. This process involves reaction of butadiene with iodine and cuprous cyanide to give the cuprous iodide complex of dehydroadiponitrile, which is further reacted with HCN to give a high yield of dehydroadiponitrile and regeneration of the iodine and cuprous cyanide. The recycling of the complexing components makes this process economically attractive.

## (c) Acrylonitrile dimerisation process

This process was developed by Monsanto and made commercial in 1965. The route is shown briefly in *Figure 2.1*. The acrylonitrile obtained from propylene is dissolved in an aqueous solution of tetraethyl ammonium p-toluene sulphonate in an electrolytic cell, the dimerisation proceeding by reduction at the cathodic electrode.

The major advantage of this process is the utilisation of cheap acrylonitrile as starting material, the disadvantage the relatively large power consumption.

## (d) Caprolactam process

This process, developed by Toray, may be used to utilise low grade lactam from nylon 6 plants and as yet is used only on a small scale. The reaction steps and conditions of operation are shown briefly in *Figure 2.1*.

## (e) Diol process

Celanese in the United States are said to have developed this process, which produces HMD from hexanediol rather than adiponitrile. Caprolactone produced as described under caprolactam (section 2.3.2.4) is hydrogenated at 250°C under 280 atmospheres using Raney nickel or copper chromite as catalysts giving a quantitative yield of the diol, which is further reacted with ammonia at about 200°C and 230 atmospheres using Raney nickel to give HMD. The yield on this final step is estimated to be 90 per cent, but a fair amount of by-product impurities is produced.

Advantages of the process are the integration of the cyclohexanone production stage with the line producing KA oil used for adipic acid. Disadvantages are the numerous steps in the process and the lower-purity diamine finally produced.

### 2.3.2 COMMERCIAL CAPROLACTAM PROCESSES

The commercial processes in current use for caprolactam, the monomer of nylon 6, are shown schematically in *Figure 2.2*[1].

The early I. G. Farben process using phenol as feedstock, first

## Raw materials for polyamides

industrialised in 1942, has been largely superseded by the method using air oxidation of cyclohexane. The explosion in the cyclohexane oxidation section at the caprolactam plant of Nypro Ltd at Flixborough, UK, in 1974 led to a reassessment of the safety aspects of the whole process. It is understood that Nypro will use a phenol feedstock in the caprolactam plant to be rebuilt on the site.

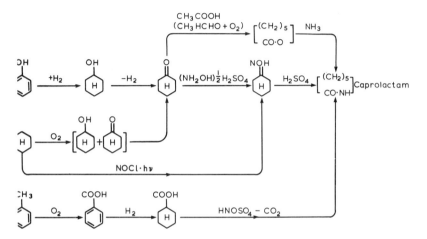

Figure 2.2  Commercial caprolactam processes (courtesy *Chemical Economy and Engineering Review*)

As will be described, more recent processes for caprolactam produce lesser amounts of ammonium sulphate by-product and some none at all.

### 2.3.2.1 Caprolactam from cyclohexanone and hydroxylamine

This is the traditional and standard route to caprolactam used by the majority of producers. Cyclohexanone and hydroxylamine are reacted together to form cyclohexanone oxime, which is then induced to undergo the Beckmann rearrangement to the lactam. In this process more or less large amounts of ammonium sulphate are produced as a by-product.

In recent years a number of innovations in the method of production of cyclohexanone and hydroxylamine have been industrialised by the main suppliers, with considerable increases in the efficiency of the conversion to caprolactam.

## (a) Production of cyclohexanone

### (i) Phenol process

This process comprises a hydrogenation stage followed by a dehydrogenation stage. In the first stage phenol and hydrogen are reacted, in the presence of a nickel catalyst, to form cyclohexanol either in the liquid or gaseous phase at a pressure below $30\,kgf/cm^2$ and a temperature of $180°C$. In the dehydrogenation stage cyclohexanol is reacted to form cyclohexanone in the gaseous phase at about $400°C$ using a copper catalyst.

Quantitative yields are obtained with the phenol process and a very pure product is obtained. The disadvantage of this process is the high price of phenol, but in the case of world shortage of alternative feedstocks for cyclohexanone this process is advantageous.

### (ii) Cyclohexane oxidation process

A mixture of cyclohexanol and cyclohexanone (KA oil) is produced by liquid-phase air oxidation of cyclohexane at a temperature of $140-180°C$ and a pressure of $10-25\,kgf/cm^2$, using a soluble cobalt salt as catalyst. The conversion rate is held below 10 per cent and recirculation of unreacted cyclohexane is required. The ratio of cyclohexanol to cyclohexanone is about $6:4$ and the reaction yield is about 75 per cent.

Research on this process by Scientific Design led to the adoption of boric acid catalysts for the oxidation, resulting in yields of up to 90 per cent. This improved process is currently used by the large suppliers including ICI, Du Pont, Rhone Poulenc, and Bayer.

The cyclohexane oxidation process is advantageous when this raw material is available at much less cost than phenol.

## (b) Production of hydroxylamine

### (i) Raschig process

This is the traditional process used by I. G. Farben in 1942 and still used widely by many suppliers of caprolactam. Ammonium nitrite is reduced with sulphur dioxide gas to give hydroxylamine disulphonate, which is decomposed into hydroxylamine sulphate. The following equation sums up the process:

$$2NH_4NO_2 + 4SO_2 + 2NH_3 + 6H_2O \rightarrow (NH_2OH)_2 \cdot H_2SO_4 + H_2SO_4 + 2(NH_4)_2SO_4$$

The process has several disadvantages. For example, two moles of ammonium sulphate are produced for each mole of hydroxylamine. Also, nitrogen oxides and sulphur dioxide must be absorbed at low temperature requiring high power costs. The process is economical where ammonia and sulphur costs are low relative to ammonium sulphate prices.

### (ii) Nitric oxide hydrogenation

This process was developed by BASF and Inventa. Pure nitric oxide is reduced with hydrogen in dilute sulphuric acid solution using platinum as catalyst. The process is represented by the following reaction:

$$2NO + 3H_2 + H_2SO_4 \rightarrow (NH_2OH)_2 \cdot H_2SO_4$$

This process has the advantage over the normal Raschig process in that only 0.5 moles of ammonium sulphate are produced for each mole of hydroxylamine after neutralisation in the oximation step. On the debit side are the recovery of the platinum catalyst and the necessity to use high-purity oxygen for the initial formation of nitric oxide.

### (iii) Reduction of nitrate ion

There is no ammonium sulphate produced in this process, developed by DSM and known as the HPO (hydroxylamine phosphate oxime) process[10]. Briefly described, the process consists of a reduction of nitrate ions by hydrogen gas in an inorganic process liquid buffered with phosphoric acid and containing a noble metal catalyst. The hydrogen ions supplied by the buffer acid for the reduction are regenerated in the oximation reaction, which is integrated with the reduction step. The buffer solution and free mineral acid are recycled after oxime separation. The process sequence is indicated by the following equations:

$$NO_3^- + 2H^+ + 3H_2 \rightarrow NH_3OH^+ + H_2O$$
$$NH_3OH^+ + (CH_2)_5C{=}O \rightarrow (CH_2)_5C{=}NOH + H^+ + H_2O$$

The oxime is recovered by solvent extraction and separation.

### (c) Oxime production and Beckmann rearrangement

Most producers still proceed through the same basic reaction steps, which are as follows. Cyclohexanone is converted into oxime by reacting the pure material with hydroxylamine sulphate. The sulphuric acid liberated in this step is neutralised using ammonia to ammonium sulphate, which is then separated from the oxime mixture. The latter in reaction with 20 per cent oleum undergoes Beckmann rearrangement, yielding the crude caprolactam mixture. After further neutralisation with ammonia the caprolactam and ammonium sulphate formed are separated by solvent extraction. Additional caprolactam is extracted from the neutralised ammonium sulphate phase, the solvent lactam phase being subjected to distillation to separate the lactam by stripping off solvent. The crude caprolactam may be purified by crystallisation or by benzene extraction followed by re-extraction using water. Other purification systems (for example, using multiple-effect evaporation of the aqueous lactam solution) are used.

The oximation and Beckmann rearrangement steps in caprolactam production are undoubtedly responsible for the production of the large amounts of ammonium sulphate by-product and are, therefore, key processes in the economic production of the intermediate.

Processes that are not dependent upon these steps are described in later sections.

### 2.3.2.2 SNIA Viscosa (benzoic acid) process

This process, patented by SNIA Viscosa in 1953, uses toluene as basic feedstock, but the distinctive feature is the reaction of hexahydrobenzoic acid and nitrosyl sulphuric acid and the avoidance of oximation and Beckmann rearrangement. The route is shown in the following series of reactions:

(1) Toluene is oxidised to benzoic acid at around 140°C in presence of a ketone and cobalt catalyst:

$$\text{C}_6\text{H}_5-\text{CH}_3 \xrightarrow[\text{Co salt}]{\text{O}_2} \text{C}_6\text{H}_5-\text{COOH} \quad \text{(93–94 per cent yield)}$$

(2) The acid is hydrogenated to hexahydrobenzoic acid using a palladium catalyst:

$$\text{C}_6\text{H}_5\text{COOH} \xrightarrow[\substack{10 \text{ atm},\\ 150°C}]{\text{H}_2} \text{C}_6\text{H}_{11}-\text{COOH}$$

(3) The acid is reacted with excess nitrosyl sulphuric acid in oleum solution at 80°C in presence of cyclohexane, yielding crude caprolactam:

$$\text{C}_6\text{H}_{11}-\text{COOH} \longrightarrow \text{caprolactam}-\text{NH} + CO_2$$

The yield of caprolactam in this process is about 90 per cent, but considerable purification of the crude product is required; about 25 per cent more ammonium sulphate is produced than with the more conventional processes using cyclohexanone and hydroxylamine.

### 2.3.2.3 PNC process

This process involves the photonitrosation of cyclohexane to yield cyclohexanone oxime hydrochloride directly. The early work was described by Lynn[4] and the patents appeared in the early 1950s.

In the process a mixture of cyclohexane nitrosyl chloride and hydrochloric acid is irradiated below room temperature with light from a mercury vapour lamp. The reactions are as follows:

$$NOCl + h\nu \longrightarrow NO^{\cdot} + Cl^{\cdot}$$

$$C_6H_{12} + Cl^{\cdot} \longrightarrow C_6H_{11}^{\cdot} + HCl$$

$$C_6H_{11}^{\cdot} + NO^{\cdot} \longrightarrow C_6H_{11}(H)(NO)$$

> 86 per cent yield: $C_6H_{10}=NOH \cdot HCl$    $HCl + h\nu$    $\lambda = 400\text{–}600\,\mu m$

$$C_6H_{10}=NOH \cdot HCl \xrightarrow[\text{Beckmann rearrangement}]{\text{oleum}} \text{caprolactam}$$

The economics of this process are related to the costs of the irradiating lamps, refrigeration costs, installation costs and the low productivity per unit. The overall efficiency of the PNC process from cyclohexane to lactam is about 80 per cent. Since the oxime has still to be oleum-treated to effect Beckmann rearrangement to caprolactam, and neutralisation by ammonia is required to release the latter, ammonium sulphate is again obtained as a by-product, but the amount is about half that produced from the traditional process using cyclohexanone and Beckmann rearrangement. Toyo Rayon (Toray) at their Nagoya plant in Japan have a full production facility based on the PNC process.

### 2.3.2.4 Lactone process

Developed by Union Carbide of the United States, two processes involving lactone have been described in patents[5]. In one process cyclohexanone and peracetic acid are reacted, in the other cyclohexanone is reacted with aldehyde and an oxygen-containing gas in the presence of a cobalt group catalyst.

This process avoids the formation of ammonium sulphate as by-product, but acetic acid almost equivalent to the lactone formed is produced as by-product. Disadvantages are the high temperatures and pressures used and the relatively low conversion rate.

## 2.3.2.5 Techni-Chem process[6]

The caprolactam process developed by the Techni-Chem Company was one of the earliest processes yielding no by-product. Up to the present time only pilot plants have been operated. The chemistry of the process is described in the series of equations set out below:

(1) Nitration (via cyclohexenyl acetate):

$$\underset{\text{cyclohexanone ketone}}{C_6H_{10}O} + CH_2CO \longrightarrow \underset{\text{cyclohexenyl acetate}}{C_6H_9OCOCH_3}$$

(in acetic anhydride)

$$\underset{\text{(in acetic anhydride)}}{C_6H_9OCOCH_3} + HNO_3 \longrightarrow \underset{\substack{\text{nitrocyclo-}\\\text{hexanone}}}{C_6H_9ONO_2} + \underset{\text{acetic acid}}{CH_3COOH}$$

(2) Cleavage of 2-nitrocyclohexanone:

$$C_6H_9ONO_2 + H_2O \longrightarrow \underset{\varepsilon\text{-nitrocaproic acid}}{O_2N(CH_2)_5COOH}$$

(3) Reduction of nitrocaproic acid:

$$O_2N(CH_2)COOH \xrightarrow{+H_2} \underset{\varepsilon\text{-aminocaproic acid}}{H_2N(CH_2)_5COOH}$$

(4) Lactam formation:

$$H_2N(CH_2)_5COOH \longrightarrow \underset{\varepsilon\text{-caprolactam}}{(CH_2)_5\text{—CO—NH}} + H_2O$$

Crude lactam is solvent-extracted from the final reaction mixture.

A comparison of the currently-operated commercial processes for caprolactam in terms of material usage, reaction conditions, advantages and disadvantages is shown in summary form in *Table 2.1*[1].

Many other processes for producing caprolactam aimed at simplifying the sequence of processes or avoiding unwanted by-products are now in active development. For example, in the DSM sulphuric acid recycle process[10] the normal ammonium sulphate by-product is pyrolysed into sulphur dioxide and nitrogen, the former being further converted to oleum and recycled; and in the Kanebo acetyl caprolactam process[11], in the final Beckmann rearrangement step, the cyclohexane-oxime is acetylated using acetyl caprolactam and the acetylated oxime subjected to Beckmann rearrangement in the

**Table 2.1** COMPARISON OF COMMERCIAL CAPROLACT...

| | | Via cyclohexanone, hydroxylamine | |
|---|---|---|---|
| | Phenol process with Raschig process | Cyclohexane oxidation process (Cobalt catalyst) with Raschig process | Cyclohexane oxidat... process (boric aci... catalyst) with Rasc... process |
| Unit consumption | (Allied process)<br>Phenol 0.89 t<br>Hydrogen 0.04 t<br>Fuming sulphuric acid 1.35 t<br>Sulphur 0.68 t<br>Ammonia 1.46 t<br>Carbon dioxide gas 0.52 t<br>Ammonium sulphate (by-product) 4.6 t | (DSM process)<br>Cyclohexane 1.06 t<br>Caustic soda 0.11 t<br>Sulphur dioxide 1.32 t<br>Ammonia 1.51 t<br>Carbon dioxide gas 0.41 t<br>Fuming sulphuric acid 1.34 t<br>Ammonium sulphate (by-product) 4.5 t | (DSM process)<br>Cyclohexane 0.9...<br>Caustic soda 0.0...<br>Sulphur dioxide 1.3...<br>Ammonia 1.5...<br>Carbon dioxide gas 0.4...<br>Fuming sulphuric acid 1.3...<br>Ammonium sulpha... (by-product) 4.5... |
| Reaction conditions and yield | *Cyclohexanone production*<br>Temperature: 200°C<br>Catalyst: Pd<br>Yield: good<br>*Hydroxylamine production*<br>Temperature ($CO_2$ and $SO_2$ absorption): <10°C<br>Yield: 70–80% (for ammonia)<br>*Beckmann rearrangement*<br>Temperature: 80–150°C<br>Yield: Almost quantitative | *Cyclohexanone production*<br>Temperature: 140–180°C<br>Pressure: 10–25 kgf/cm$^2$<br>Catalyst: cobalt salt<br>Reaction time: ≈1 h<br>Conversion rate: <10%<br>Yield: 70–75% | *Cyclohexanone production*<br>Catalyst: metabo... a...<br>Reaction time: ≈2...<br>Conversion rate: ≈1...<br>Yield: ≈9...<br>Temperature and pressure approxim... those for cobalt catalyst process |
| Advantages | Good yield<br>Purification easy<br>Reaction easy | Lower capital cost than boric acid process<br>Reaction easy<br>Good yield, except cyclohexanone | Overlapping s...<br>All steps have good yield<br>Reaction easy<br>Purification easy |
| Weaknesses | High cost of phenol, and hydroxylamine<br>Numerous steps<br>Much by-product ammonium sulphate | Low cyclohexanone yield<br>Hydroxylamine costly<br>Much by-product ammonium sulphate | Cyclohexanone pla... costly<br>Hydroxylamine cos...<br>Much by-product ammonium sulph... |

...OCESSES (courtesy *Chemical Economy and Engineering Review*)

| ...lohexane oxidation process (Cobalt catalyst) with NO ...eduction process | PNC process | Benzoic acid process |
|---|---|---|
| Inventa process) | | |
| ...clohexane 1.1 t | Cyclohexane 0.93 t | Toluene 1.11 t |
| ...ustic soda 0.1 t | Ammonia 0.82 t | Cyclohexane 0.025 t |
| ...monia 1.0 t | Fuming sulphuric | Ammonia 1.25 t |
| ...ming sulphuric | acid, sulphuric | Hydrogen 0.08 t |
| ...cid, sulphuric | acid 1.72 t | Fuming sulphuric |
| ...cid 1.8 t | Ammonium sulphate | acid, sulphuric |
| ...ygen 400 m³ | (by-product) 2.29 t | acid 2.90 t |
| ...drogen 500 m³ | Chlorocyclohexane | Caustic soda 0.16 t |
| ...monium sulphate | (by-product) 0.064 t | Ammonium sulphate |
| (by-product) 2.6 t | (Electric power | (by-product) 4.1 t |
| | 3950 kWh) | |
| ...droxylamine ...duction ...nditions for NO ...uction) ...emperature: 30–80°C ...ressure: 1–8 kgf/cm² ...atalyst: Pt ...ield: ≈ 85% (for NO) | Nitrosation reaction Temperature: 10–20°C Yield: > 86 wt% Oxime yield per unit power consumption: 330 g/kWh | BA production Temperature: 150–170°C Yield: 93–94% HHBA production Temperature: ≈ 150°C Pressure: ≈ 10 kgf/cm² Yield: almost quantitative Lactam production Temperature: ≈ 80°C Yield for HHBA: 95–96% Yield for N: 90% |

| ...ted | | |
|---|---|---|
| ...wer capital cost ...an boric acid ...ocess ...duced by-product ...mmonium sulphate | Lessened reaction steps Reduced by-product ammonium sulphate | Uses toluene as raw material |
| ...mplex operating ...nditions for ...droxylamine step ...w yield of ...clohexanone | Equipment of special material required Large power requirements Nitrosation reactor offers little scale up merit | Severe conditions for lactam preparing reaction Much by-product ammonium sulphate Purification difficult |

vapour phase, thereby dispensing with oleum and avoiding sulphate by-product at this stage.

### 2.3.3 LAURINLACTAM

Considerable research has been carried out on other lactams and in the last few years laurinlactam (dodecalactam), the intermediate for nylon 12, has been produced commercially in Germany by Hüls and in France by Aquitaine Organico. The Japanese companies of Mitsubishi and Toray in cooperation with other companies have planned, or are about to start up, units for laurinlactam production[12].

The route from the feedstock butadiene to laurinlactam is rather involved. The following equations describe briefly one commercial method:

(1) Butadiene is trimerised over a Ziegler catalyst to cyclododecatriene 1, 5, 9 (CDT):

$3H_2C=CH-CH=CH_2 \longrightarrow$ CDT

(2) The CDT is hydrogenated to cyclododecane (CD):

$\xrightarrow{H_2}$ CD

(3) The CD is reacted with nitrosyl sulphuric acid photochemically using mercury vapour lamps to yield the oxime:

$+ H_2SO_4 \cdot NO \longrightarrow$ —NOH

(4) The product is treated with oleum and the laurinlactam formed by Beckmann rearrangement:

—NOH $\xrightarrow{\text{oleum}}$ CO NH  laurinlactam

Other models for the conversion of cyclododecane to the lactam are described in the patent literature[8].

## REFERENCES

1. Nawata, G., *Chemical Economy and Engineering Review* (May 1972)
2. *Chemical Week* (12 May 1971) and US Patents 3 526 654, 3 547 972 (1970)
3. SNIA Viscosa, Italian Patents 603 606 and 604 795 (1958)
4. Lynn, E. V., *J. Am. chem. Soc.*, **41**, 368 (1919); **45**, 1045 (1923)
5. Japanese Patents 34-5663, 39-5921
6. Caprolactam Supplement, 28, *European Chemical News* (2 May 1969)
7. *European Chemical News*, 31 (14 May 1971)
8. Badische Anilin und Soda Fabrik, German Patent 1 079 036
9. *European Chemical News*, 20 (9 Feb. 1973)
10. *European Chemical News*, 21 (22/29 Dec. 1972)
11. *European Chemical News*, 35 (11 May 1973)
12. *Chemical Age International* (6 Feb. 1973)

# 3
# Basic chemistry and manufacture of polyamides

## 3.1 Classification and nomenclature

A description of the synthetic methods and chemical routes to the many polyamides now commercially available is given below in sufficient detail for an appreciation of their technical manufacture, which will be described later in the chapter.

Three broad classes of polyamides can be distinguished according to the basic mechanism involved in the build-up of the polymer from the monomeric units. These are:

(1) polyamides formed directly by stepwise polymerisation,
(2) polyamides formed through initial ring scission reactions,
(3) polyamides formed by synthetic methods other than those of 1 and 2.

Physico-chemical considerations may also be used to classify the synthetic route, viz.

(1) melt polymerisation,
(2) solid-state polymerisation,
(3) low-temperature polymerisation:
 (a) interfacial polymerisation,
 (b) solution polymerisation.

A further classification concerns the recurring unit in the polymer. The class of homopolyamides has only one species as recurring unit, while copolyamides have two or more recurrent units in an irregular or regular sequence along the polymer chain.

Examples of the classes and types of polyamide will be described later in the chapter. It is convenient, in order to appreciate the make-up of the polymers from the basic units or monomers, to use a conventional shorthand system where A denotes an amine group

## Basic chemistry and manufacture

and B a carboxyl group. A monomer, for example ε-aminocaproic acid, that has two different functional groups (in this case an amine and a carboxyl) may be represented by the symbol AB. Polyamides derived from AB monomers are known as type AB polyamides.

Monomers may also comprise a diamine component, for example hexamethylene diamine, which can be represented by the symbol AA, and a dicarboxylic acid component, for example adipic acid, represented by BB. Polyamides derived from these components are termed AA BB type.

As mentioned in section 2.1, individual polyamides may also be identified by a number code. For an AB type polymer a single number is used, numerically the same as the number of carbon atoms in the monomer. For an AA BB type polymer two digits are normally used, the first giving the number of carbon atoms separating the nitrogen atoms of the diamine, the second the number of straight-chain carbon atoms in the dibasic acid.

Ring-containing monomer components are usually coded with single letters or short letter combinations to denote the grouping—for example, T for terephthalic acid and Pip for piperazine.

From the foregoing it may be seen that linear polyamides formed from aminoacids or lactams, i.e. AB type, may be represented by a structure such as

$$A-BA-BA-BA-BA-B$$

where the dash between functional groups represents an alkyl chain segment. The recurring structural unit in this case is of the same composition as the monomer. The number of units forming the complete chain is generally termed the degree of polymerisation, and in the example given this is five.

Polyamides formed from diamines and dicarboxylic acids, i.e. the AA BB type, may be represented by a structure such as

$$A-AB-B/A-AB-B/A-AB-B/A-AB-B$$

Here the recurring unit, the monomer, is the grouping A—AB—B, and the degree of polymerisation in this example is four.

### 3.2 Direct stepwise polymerisation

For polyamides this type of reaction is exemplified by the condensation of a dibasic acid and a diamine with elimination of water during the polymerisation. Type AA BB polyamides as described in the previous paragraph are produced. The only two polyamides of commercial importance in this class are nylon 66 and nylon 6.10. Their technical manufacture is described below.

## 3.2.1 NYLON 66—TECHNICAL MANUFACTURE

The manufacture of this most important of all the polyamides uses as starting intermediate 'nylon salt', which is formed technically by reaction of a 60–80 per cent methanol solution of hexamethylene diamine with a 20 per cent methanol solution of adipic acid at a temperature of 50°C and a pH of 7.6. The reaction is in accordance with the following equation:

$$nNH_2(CH_2)_6 \cdot NH_2 + nHOOC(CH_2)_4 \cdot COOH$$
$$\downarrow$$
$$nNH_2(CH_2)_6NH_2 \cdot HOOC(CH_2)_4COOH$$

The salt is sparingly soluble in methanol, and after cold washing with the solvent and drying it can be recovered in a very pure form. It melts between 202°C and 205°C. The solvent is caused to boil due to the heat of neutralisation during process and it can be recovered and recycled.

To effect neutralisation it is important to use strictly a stoichiometrical proportion of diacid and diamine.

Condensation of the 66 salt to the polymer is carried out under pressure in a stainless steel autoclave. A 60 per cent aqueous solution of the 66 salt to which end-group stabiliser (commonly acetic acid) has been added is pumped into the autoclave, which is then completely purged with oxygen-free nitrogen and sealed. The temperature of the contents is slowly raised to about 210°C by circulating heat-transfer fluid through the jacket. The azeotrope of diphenyl and diphenyl oxide (Dowtherm), which has a boiling point under standard conditions of 265°C, is normally used for this purpose. The temperature of the reactants is maintained at 210°C and the pressure at 17.5 kgf/cm$^2$ (250 lbf/in$^2$) for several hours, during which time initial condensation occurs although the reaction mixture remains quite fluid and low-molecular-weight products remain dissolved. The temperature is then further raised slowly to 275°C, steam being bled off to maintain a pressure of about 17 atmospheres. The liquid reaction products become more and more viscous as the condensation proceeds. Thereafter pressure is slowly released over a period of one hour while heat is supplied to the autoclave jacket to maintain the temperature.

Stirring of the viscous mass of the reactants is effected by ebullition from steam bubbles arising from the condensation reaction. The condensation phase must, however, be kept as short as possible to prevent degradation of the reactants, and as soon as the reaction has gone to completion the melt is extruded through a strip die in the base of the autoclave on to a water-cooled roller. Further cooling of the strip is effected by water and air jets, and the strip is then fed to a rotary cutting machine where it is comminuted to granules of a size suitable for further reprocessing. *Figure 3.1*[1] shows diagrammatically the equipment for manufacturing nylon granules from 66 salt.

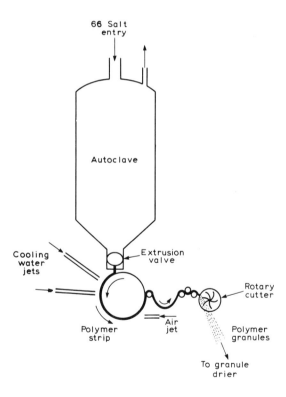

Figure 3.1  Diagram of equipment for manufacturing nylon 66 granules (courtesy Carl Hanser Verlag)

The granules may be in the form of plate-like 'chips' if required for subsequent melt spinning, or cylindrical granules if used for extrusion and injection-moulding. Drying may consist of surface drying of the comminuted material to remove water in excess of the equilibrium amount determined by the final equilibrium conditions in the autoclave; alternatively tumbler-drying under vacuum can be used to reduce the moisture content of the nylon even further.

In the discontinuous process described above small variations in the characteristics of each batch are averaged out by batch bulking. To overcome this disadvantage several continuous methods for 66 salt polycondensation have been proposed and are described in the patent literature[2].

In all these continuous processes it is essential to avoid loss of diamine, so the reactants are usually added slowly to a much larger quantity of partially condensed material. Loss of heat due to

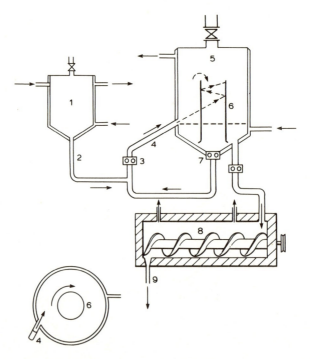

Figure 3.2 Diagram of continuous polymerisation of 66 salt; see text for description (courtesy Carl Hanser Verlag)

## Basic chemistry and manufacture

evaporation of water of condensation in the later stages of reaction must also be avoided to obviate precipitation of the solid polymer. The arrangement proposed by NV Onderzoekinginstitut of Holland is shown in *Figure 3.2* and described below.

A 60 per cent aqueous solution of 66 salt stabilised with 0.5 mol per cent of acetic acid is led into the jacketed hold tank (1), containing partially polymerised material. This tank is maintained at 250°C and 18 atmospheres pressure. In a residence time of about one hour the reactants remain in the tank and are delivered slowly through the line (2) into the circulating line (4), which already contains polymerised reactants from the vessel (5), held at a lower pressure than (1). New material entering through (2) is held inside (5) by the action of the gear-pump (3), which runs 11 per cent faster than the gear pump (7). Thus for every ten volumes flowing from vessel (5) only one volume enters from vessel (1).

Mixing is effected by means of the spiral motion of the flow around cylinder (6). The temperature inside (5) is maintained at 275°C by the heating jacket, and the pressure controlled at atmospheric by bleeding off the steam formed. The polymerised mixture is pumped from this vessel to the final condensation chamber (8), fitted with a screw conveyor. Here more steam is released and the final reaction product is delivered through (9).

### 3.2.2  NYLON 6.10

Like nylon 66, nylon 6.10 is prepared through the intermediate of a nylon salt. The 6.10 salt crystallises well with a melting point of 170°C.

The technical preparation of the polymer follows a route very similar to that of nylon 66. The higher stability of the sebacic acid component results in less degradation, and a longer polymerisation time can be tolerated than in the case of nylon 66.

### 3.3  Ring scission polymerisation—
### hydrolytic and anionic polymerisation

This type of reaction mechanism is characteristic of lactams. The ring opening step is induced either catalytically or in the presence of a small amount of water; in the latter case the mechanism is termed hydrolytic polymerisation. On the other hand, lactams can also be polymerised by an ionic chain mechanism where the monomer attaches itself to active centres bearing negative charges. This mechanism, termed anionic polymerisation, differs in many essential features from hydrolytic polymerisation. Both mechanisms are considered in detail in this chapter.

As a rule lactams with six or more carbon atoms can be polymerised by either route, while smaller ring lactams must be polymerised by ionic methods. These may be anionic or cationic. Considerable work on elucidating the mechanisms of these processes was carried out by Hermans *et al.*[3] and Wiloth[4]. The currently accepted kinetic schemes have been summarised by Reimschüssel[5]. Since caprolactam is by far the most technically important of the lactams, the mechanisms described in the following sections will refer exclusively to this compound, as typical of the mechanism.

### 3.3.1 HYDROLYTIC POLYMERISATION–CHEMISTRY

Hermans showed that highly purified caprolactam could be held for a prolonged period at 250°C without undergoing polymerisation even in the presence of organic acids or bases. On the other hand, 66 salt and other compounds capable of yielding free water at elevated temperatures could considerably accelerate the reaction leading to polymerisation. The generally accepted mechanism comprises three steps, shown in the following equations:

(1) *Ring Opening*

$$NH(CH_2)_5CO + H_2O \rightleftharpoons H_2N(CH_2)_5COOH$$

(2) *Polycondensation*

$$H_2N(CH_2)_5COOH + H_2N(CH_2)_5COOH$$
$$\updownarrow$$
$$H_2N(CH_2)_5CO\!-\!NH(CH_2)_5COOH + H_2O$$

Dimers and oligomers with free amine and carboxyl ends can also act as monomers in this reaction.

(3) *Polyaddition*

$$NH_2(CH_2)_5COOH + \overline{HN(CH_2)_5CO} \rightleftharpoons H_2N(CH_2)_5CO$$
$$HOOC(CH_2)_5NH$$

Cyclic oligomers can also act as monomers in this reaction.

Using end-group analysis and chromatographic methods for determining free amino acid concentrations, it was found possible to establish the kinetic processes. Plots of the concentration of the main species against reaction time revealed that the polymerisation reactions were catalysed by end-groups and that the predominant reaction was polyaddition.

Two further reactions occurring during polymerisation were studied later. These were the uncatalysed reaction between lactam and aminocaproic acid and the end-group catalysed transamidation reaction between lactam and an amide group of the linear chain, but neither of these appreciably altered the predominant polyaddition reaction.

In brief, the mechanism of the hydrolytic polymerisation of caprolactam would seem to be a ring-opening hydrolysis of the lactam catalysed by carboxyl groups, followed by a lactam addition to the amino end-groups of the linear chain similarly catalysed.

## 3.3.2 ANIONIC POLYMERISATION–CHEMISTRY

The anionic polymerisation of caprolactam is a base-catalysed reaction. Using suitable initiators, conversion to the equilibrium state can be brought about within minutes, which makes the process attractive from the commercial viewpoint.

Considerable research has been carried out in studying the process mechanism after the first description by Joyce et al[6]. The currently accepted mechanism is that proposed by Mottus et al[7] and, independently, by Sebenda and Kralicek[8]. It is shown that when caprolactam is heated in presence of a base, for example sodium caprolactam, polymerisation proceeds with increasing velocity after a certain induction period. It is proposed that active amino and imide groups are formed in the initial stages, but since amine groups were shown to inhibit rather than initiate polymerisation, the active initiating group is identified as the imide. Polymerisation is therefore considered to be a nucleophilic attack of the caprolactam anion on the carboxyl of the imide group. The following equations show the main steps in the reaction mechanism.

*Catalyst formation*

$$(CH_2)_5\!-\!CO\!-\!NH \xrightarrow{NaH} (CH_2)_5\!-\!CO\!-\!N^{\ominus} + H_2\uparrow \quad \text{lactam ion} \quad (3.1)$$

*Normal initiation*

$$(CH_2)_5-\underset{|}{\underset{NH}{C}}O + (CH_2)_5-\underset{|}{\underset{N^\ominus}{C}}O \rightarrow (CH_2)_5-\underset{|}{\underset{\substack{N \\ | \\ CO-(CH_2)_5-NH^\ominus}}{C}}O \quad (3.2)$$

*Activated initiation*

$$(CH_2)_5-\underset{|}{\underset{\substack{NCH_3CO \\ \text{acetyl lactam}}}{C}}O + (CH_2)_5\underset{|}{\underset{N^\ominus}{C}}O \xrightarrow{\text{fast}} \underset{|}{\underset{\substack{N-(CH_2)_5CON-(CH_2)_5 \\ | \\ CO}}{CH_3CO}} \quad (3.3)$$

$$\left.\begin{array}{l}
\underset{|}{\underset{\substack{N-(CH_2)_5CO\cdot N-(CH_2)_5 + (CH_2)_5CO \\ \ominus \quad\quad | \quad\quad\quad\quad\quad | \\ CO \quad\quad\quad\quad\quad NH}}{CH_3CO}} \\
\downarrow \text{proton exchange} \\
\underset{|}{\underset{\substack{NH-(CH_2)_5-CO\cdot N-(CH_2)_5 + (CH_2)_5-CO \\ | \quad\quad\quad\quad\quad\quad | \\ CO \quad\quad\quad\quad\quad\quad N^\ominus}}{CH_3CO}}
\end{array}\right\} \quad (3.4)$$

*Propagation*

$$\left.\begin{array}{l}
CH_3CO-[NH-(CH_2)_5-CO]_1-\underset{|}{\underset{CO}{N}}-(CH_2)_5 + (CH_2)_5-\underset{|}{\underset{N^\ominus}{C}}O \\
\downarrow \\
CH_3CO-[NH-(CH_2)_5-CO]_1-\underset{\ominus}{N}-(CH_2)_5-CO-\underset{|}{\underset{CO}{N}}-(CH_2)_5
\end{array}\right\} \quad (3.5)$$

$$\left.\begin{array}{l}
CH_3CO-[NH-(CH_2)_5-CO]_1-\underset{\ominus}{N}-(CH_2)_5-CO-\underset{|}{\underset{CO}{N}}-(CH_2)_5 + (CH_2)_5-\underset{|}{\underset{NH}{C}}O \\
\downarrow \\
CH_3CO-[NH-(CH_2)_5-CO]_2-\underset{|}{\underset{CO}{N}}-(CH_2)_5 + (CH_2)_5-\underset{|}{\underset{N^\ominus}{C}}O
\end{array}\right\} \quad (3.6)$$

The slow normal initiation as shown in equation (3.2) can be completely eliminated by using initiators with imide groupings such as acyl lactams. Conversion proceeds fast, as shown in equations (3.3) and (3.4), and it has been shown that in the presence of both a

## Basic chemistry and manufacture

catalyst and an imide initiator bulk polymerisation can be carried out between 100°C and 200°C, i.e. below the melting point of the polymer, which is around 225°C. The propagation steps appropriate to the activated initiation are shown in equations (3.5) and (3.6).

The polymerisation of caprolactam is an exothermic process and, in the case of activated anionic polymerisation with rapid conversion, the conditions are adiabatic. The high viscosity and low heat conductivity result in a considerable rise in temperature in the polymerising mass, and the rate of this increase can be considered a measure of the rate of the reaction.

The heat of polymerisation of a lactam depends on the ring size. $Table\ 3.1$[9] shows $-\Delta H$ liquid lactam to liquid polymer for several of the species.

Table 3.1  HEAT OF POLYMERISATION OF LACTAMS
(courtesy *Angewante Chemie*)

| Monomer | $-\Delta H$ kcal/mol |
|---|---|
| Caprolactam | 3.8 |
| Enantholactam | 5.2 |
| Capryllactam | 9.6 (approx.) |
| Laurinlactam | 1.4 (approx.) |

It has been observed that for polymerisation below the melt the equilibrium monomer is appreciably less than calculated from theoretical considerations. This effect has been explained by the presence of a crystalline and an amorphous phase, the former containing no monomer and not participating in the polymer-monomer equilibrium.

For below-the-melt polymerisation the entropy changes arising from recrystallisation also have to be taken into account, and the exotherm will vary with the particular lactam.

### 3.3.3 THE HYDROLYTIC PROCESS— NYLON 6 TECHNICAL MANUFACTURE

The basic features of the hydrolytic process are as follows. The lactam in 20 per cent aqueous solution, to which has been added a chain terminator (usually acetic acid), is heated in a closed reactor to 260–270°C during which phase free amino acid is formed by hydrolysis. On reaching temperature, pressure is slowly reduced to atmospheric by bleeding off steam and a further 3–4 hours is allowed for further condensation to polymer. Since stabiliser is present the reaction may proceed to equilibrium. Oxygen must be strictly excluded from the reaction vessel to prevent oxidative degradation, which shows up as a yellowing of the polymer.

The batch process is virtually two-stage, the first carried out at high water content and pressure, the second at a lower water content at around atmospheric pressure. In the latter phase time taken to reach equilibrium can be long and depends upon the final water content of the polymer—e.g. at 260°C, 10 hours for 1 per cent of water and 36 hours for 0.25 per cent.

Several factors affect the kinetics and equilibrium during the initial stages of melt polymerisation: for example, the initial water content of the mix, and the temperature and pressure during polymerisation. All these must be rigidly controlled to ensure uniformity of the product. The relationship between polymerisation time to equilibrium and temperature of melt for different initial water contents is shown in *Figure 3.3*[11].

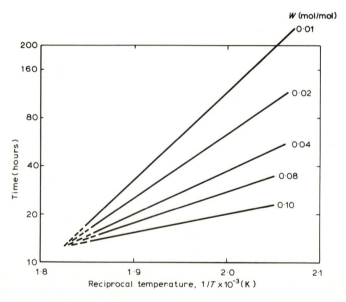

Figure 3.3 Hydrolytic nylon 6: time to attain equilibrium vs temperature for various initial water contents, $W$ (courtesy *Journal of Polymer Science*)

## Basic chemistry and manufacture

From the kinetic and equilibrium data available, von Krevelen and co-workers[12] worked out the conditions for optimising the continuous hydrolytic polymerisation (see below) of 6-polyamide. They showed that removal of water between the high and low pressure stages should be carried out as quickly as possible.

While batch production of hydrolytic 6-polyamide can produce high-grade material, a more uniform product is generally produced by the continuous process first established in Germany by Wenger and Ludewig[13] and subsequently protected by a large number of patents. This particular process is called the VK Process. A single reaction tube arrangement is shown schematically in *Figure 3.4* and the steps in the process are as follows.

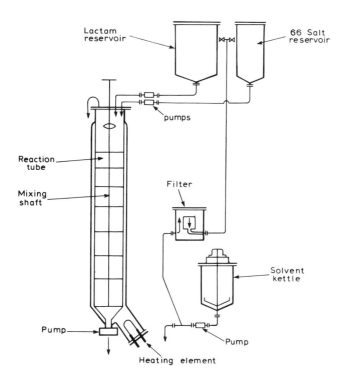

Figure 3.4 Diagram of VK Process for continuous hydrolytic polymerisation of nylon 6 (courtesy Carl Hanser Verlag)

The aqueous monomer together with the chain terminator is delivered to the upper section of the 4–6 metre long pre-heated reaction tube. Crude polymer issues from the lower end of the tube. For starting up the run, dry lactam, chain terminator and initiator (e.g. 66 salt) are added only. When the run is under way aqueous lactam can be added, because by that time sufficient amino acid has been produced to initiate chain formation. The temperature is held at about 265°C during the run, but the equilibrium moisture content reduces with temperature and it is advantageous to keep this as low as possible. Generally polymer issuing from the tube contains 0.2–0.4 per cent water. Throughput can be increased by using wider-diameter tubes, but this is at the expense of molecular weight, which is reduced in the final product unless the path length of the material through the reactor is increased by using baffles and perforated plates. The latter also serve to distribute the rising steam bubbles and prevent locks.

6-polyamide produced by both batch and continuous processes contains an unacceptable proportion of monomer and oligomers if the material is intended for plastics applications. This is due to the tendency to attain equilibrium during processing, which under the conditions obtaining favours an appreciable content of low molecular weight material. Removal of these low molecular weight compounds is simply effected by repeated water extraction, sometimes using a proportion of added reducing agent to prevent colour degradation of the nylon granules. If water extraction is used a further drying step is required to remove water.

Monomer and oligomers may alternatively be removed by heating the granules in nitrogen, when lactam is released at about 200°C. The system must be fully stabilised to prevent further condensation. Polymer prepared in this way has very low moisture content, and for further processing (for example, injection moulding) it must be sealed in air-tight containers to maintain its dryness prior to these operations. Thin-film techniques and vacuum drying have been proposed to speed up the drying process, but care must be taken to ensure the polymerisation proceeds no further. Superheated steam can be used to remove low molecular weight residues without incurring risk of further polymerisation.

It is important to appreciate that, during any later conversion processing of the extracted nylon polymer, there is a tendency for the equilibrium to be re-established. Conditions used should, therefore, be such that the tendency to form low molecular weight products is minimised.

## Basic chemistry and manufacture

### 3.4 Other stepwise polymerisation mechanisms

In addition to the methods described in previous sections, others have been the subject of patents[14] but have not been commercialised. For example, the reaction of nitriles with olefines and tertiary alcohols can take place according to the following equations ($R''$ = alkylene radical):

$$NC-R-CN + HO\overset{|}{\underset{|}{C}}-R''-\overset{|}{\underset{|}{C}}-OH$$

$$\downarrow \text{strong acid}$$

$$NC-R''-CONH\overset{|}{\underset{|}{C}}-R''-\overset{|}{\underset{|}{C}}-OH$$

### 3.5 Other commercial nylons—mechanisms and technical manufacture

#### 3.5.1 NYLON 11

This polyamide is manufactured essentially by amino-acid condensation, and was first prepared by Carothers in 1935. Today the bulk of the manufacture is undertaken by the firm of Aquitaine Organico in France, where it is marketed under the trade name *Rilsan*. The monomer, ε-aminoundecanoic acid, is derived from castor oil through ricinoleic acid, its principal ingredient. Polymerisation of the amino acid is carried out by melt condensation at 215°C under nitrogen, and the process can be made continuous.

In the technical manufacture an aqueous suspension of the monomer together with stabilisers and other additives is fed to a vertical tubular reactor. Condensation proceeds with rise in temperature, and excess water is vaporised in the upper section of the reactor. Equilibrium conditions are reached in the lower section of the reactor on completion of condensation.

The thermal stability of the polymer in the melt make it an attractive candidate for extrusions and mouldings.

#### 3.5.2 NYLON 12

Nylon 12 is now competing strongly with nylon 11 as a commercial product. The properties of the two types are very similar but as the price of butadiene, the source of the laurinlactam monomer, is likely to be more stable than that of castor oil, the precursor of nylon 11, the latter is commercially at a disadvantage.

Nylon 12 was introduced commercially in Germany by the firm of Chemische Werke Hüls AG and marketed under the trade name Vestamid. Dr Plate GmbH in Germany and Aquitaine Organico in France also manufacture the polymer, which is now freely available in the United States as well as in Western Europe.

The polymer is produced from laurinlactam, which is formed from butadiene according to the scheme described in section 2.3.3. As the 12-carbon atom laurinlactam ring is not in a state of strain there is little tendency to ring scission under the action of water as is the case with lactams containing fewer carbon atoms in the ring. The low solubility of water in the polymerising melt further reduces the tendency to ring scission.

While the polyaddition reaction occurs slowly in presence of the free amino acid it is considerably accelerated by acids used as molecular-weight stabilisers.

A detailed review of synthetic methods for laurinlactam is given by Franke and Müller[15].

### 3.5.3 NYLON 7

This polyamide (poly $\omega$-enanthamide) is commercially available in the USSR, where it is marketed under the name *Enant*. Among its very useful properties are its very low monomer content, high chemical stability and fairly high melting point (253°C). The polymerisation process is rapid and can be made continuous. The process itself, which is described by Nesmejanow and co-workers[16], is essentially a melt condensation of 7-aminoheptanoic acid. A nitrogen pressure of 15 atmospheres and a temperature of 250°C are used.

The monomer can be prepared by at least two methods. The first is the process known as 'telomerisation' (which may be defined as a method of intentionally terminating polymer chain growth by a radical donating agent). Here ethylene is led into carbon tetrachloride containing the agent and telomerises under 100 atmospheres pressure. This is followed by carboxylation of the trichlormethyl group and ammination of the $\omega$-chlorheptanoic acid ($\omega$-chlorenanthic acid) to yield the amino acid. The sequence is represented by the following equation and the process is described in detail in the literature[17].

## Basic chemistry and manufacture

$$CCl_4 + 3(CH_2{=}CH_2) = Cl(CH_2)_6CCl_3$$

$$Cl(CH_2)_6CCl_3 \xrightarrow{H_2SO_4} Cl(CH_2)_6COOH$$

$$\downarrow NH_3$$

$$NH_2(CH_2)_6COOH + NH_4Cl$$

In practice, mixed 'telomers' result from the reaction due to the random nature of the chain growth, and mixtures having six, eight and ten carbon atoms in the alkyl chain are possible. To recover individual species the mix has to be fractionated usually under reduced pressure.

In the second method, the monomer is prepared from caprolactone by ring opening with hydrochloric acid, followed by esterification, cyanation and reduction over a nickel catalyst. This sequence produces the amino-acid ester according to the following equations:

$$O(CH_2)_5C{=}O \xrightarrow[ZnCl_2]{HCl} Cl(CH_2)_5COOH$$

$$\downarrow \begin{array}{c}ROH\\H^+\end{array}$$

$$Cl(CH_2)_5COOR \xrightarrow{NaCN} NC(CH_2)_5COOR$$

$$\downarrow \begin{array}{c}H_2\\Ni\end{array}$$

$$NH_2(CH_2)_6COOR$$

Polycondensation of the ester of heptanoic acid is carried out at 90–100°C in excess water. The reaction is complete in a few hours, but is considerably slower than the polycondensation of the free amino acid. The lactone route for nylon 7 is, however, technically to be preferred.

### 3.5.4 NYLON 8

While nylon 8 is still not commercially available the basic intermediates for preparation of the monomer are in ready supply, and in the fullness of time this type might well become viable.

The route to the final product is interesting. The monomer,

capryllactam, can be obtained by two methods, which proceed via cyclooctane. These are illustrated below:

(1) Catalytic cyclisation of acetylene to cyclooctatetraene:

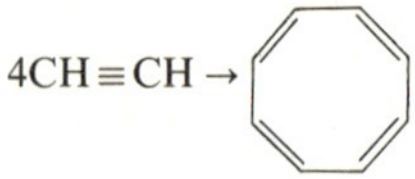

$$4CH \equiv CH \rightarrow \bigcirc$$

(2) Catalytic cyclisation of butadiene[19] to cyclooctadiene:

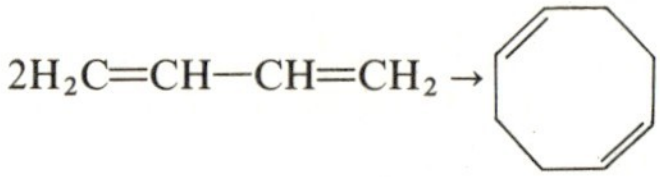

$$2H_2C=CH-CH=CH_2 \rightarrow \bigcirc$$

The two unsaturated eight-member rings can be hydrogenated to cyclooctane, which can then be converted to the corresponding lactam by the variety of methods much in the same way as cyclohexane to caprolactam.

Hydrolytic polymerisation of capryllactam proceeds readily in the melt faster than in the case of caprolactam, and provision must be made to remove the heat developed during the reaction to avoid vaporisation of monomer. Anionic polymerisation of the lactam is also possible, and here again reaction is rapid. The heat of reaction is considerably greater than for caprolactam.

Nylon 8 has properties approximately intermediate to those of nylon 6 and nylon 11, with good thermal stability. Monomer content is generally low.

### 3.5.5 NYLONS FROM POLYMERISED VEGETABLE OILS

These nylons are an important, relatively new, class commercialised widely in the US and produced under licence in the UK and elsewhere. They are liquid or low melting, highly soluble polyamide resins of low to moderate molecular weight, produced from the condensation of diamines or triamines with relatively high molecular weight dibasic acids or esters obtained from thermal polymerisation of a diene acid or ester such as linoleic acid. A typical thermal reaction is as follows[20]:

$$\begin{array}{c}COOCH_3\\(CH_2)_7\\CH\\\|\\CH\\|\\CH\\\|\\CH\\|\\(CH_2)_5\\|\\CH_3\end{array} + \begin{array}{c}COOCH_3\\(CH_2)_7\\CH\\\|\\CH\\|\\CH_2\\|\\CH\\\|\\CH\\|\\(CH_2)_4\\|\\CH_3\end{array} = \begin{array}{c}COOH_3\\(CH_2)_7\\CH\\\diagup\;\diagdown\\HC\quad CH-(CH_2)_7-COOCH_3\\\|\quad\;\,|\\HC\quad CH-(CH_2)-CH=CH-(CH_2)_4-CH_3\\\diagdown\;\diagup\\CH\\|\\(CH_2)_5\\|\\CH_3\end{array}$$

The raw linoleates are derived from soyabean, cottonseed or corn oil.

In the technical manufacture the reactants are blended at low temperature in a stirred autoclave. On gradual heating, water is formed and distilled off continuously at 100°C. Towards the end of the condensation the temperature rises to around 200°C and the reaction is completed under vacuum, after which inert gas is introduced to break the vacuum and allow discharge of the product.

Nylons from polymerised vegetable oils generally form two series:

(1) The 'solid' series, whose members are used alone or modified with plasticisers or waxes. The products are used as hot melt cements, sealants, and so on.
(2) The 'reactive' series, which may be used to modify members of the solid series or may be converted to thermoset resins by reaction with epoxy resins, phenolic resins and other similar resins.

Other types are possible, using isophthalic or terephthalic acids, or aromatic diamines, but not all have been commercialised.

## 3.6 Polyamides from aromatic diamines

Although not commercially available, polyamides derived from aromatic diamines have been the subject of intensive development and a host of patents has resulted. A detailed review of this field up to the year 1965 is given by Morgan[21].

From the chemical viewpoint the processes for production of aromatic polyamides are interesting, in that use has to be made of the techniques of interfacial or solution polycondensation (see sections 3.9.3.1 and 3.9.3.2), melt condensation being unsuitable because of the low reactivity of the amines and the possibility of condensation at the high melt temperature.

For interfacial polymerisation of polyamides it seems essential, for good yields, to use a solute with good water solubility, water being the other phase in the two-phase mix. Solution polymerisation methods seem preferable for the preparation of aromatic polyamides, and high yields of high molecular weight material have been reported.

## 3.7 Copolyamides

Copolymerisation is defined in IUPAC (International Union of Pure and Applied Chemistry) Information Bulletin No. 13 as the polymerisation of a mixture of two or more species of monomers.

In the simultaneous polymerisation of two or more monomers in a system undergoing stepwise polymerisation the reactivity of the functional groups are, in theory, equal irrespective of the length of the molecule to which they are attached. The comonomers are, therefore, randomly distributed along the chain in amounts proportional to their concentration in the original mix. This ideal situation may be modified by the steric effect of one of the monomers, which might, for example, have bulky side groups that reduce the reactivity of the adjacent functional group.

Copolyamides may be prepared by polymerisation of the mixed monomers or by blending the molten polymers using continuous heating for varying periods. In the latter case the copolymerisation process must proceed through exchange reactions.

The identity of copolymers produced from mixtures of nylon 66 and nylon 6.10 and separately, from mixed monomers of the same composition, has been demonstrated using chromatographic analysis. Melt point and solubility of compositions thus shown to be identical were also the same.

Structurally two types of copolyamides can be distinguished:

(1) The two components form isomorphous crystals.
(2) The two components are anisomorphous.

## Basic chemistry and manufacture

For the first type the properties of the copolymers are approximately intermediate between those of the two homopolymers (in the case of a two-component system) and can be forecast from a knowledge of the relative proportions of the comonomers in the system. Properties of copolymers of the second type differ substantially from those of either of the homopolymers concerned.

An example of the first type is given by the system nylon 66/6T (6T = hexamethylene terephthalamide) studied by Edgar and Hill[22], which is shown in *Figure 3.5*. The graph shows that for the isomorphous system there is no composition of copolymer with a melting point less than that of the lower melting homopolymer. The curve relating the melting point to composition is generally S-shaped. The explanation of this fact is that the copolymerising segments of the separate acidic groups of the two comonomers (in this case adipic acid and terephthalic acid) are similar in size, and the symmetry and order along the chain are relatively undisturbed if one type of segment replaces the other. Thus crystallinity and the

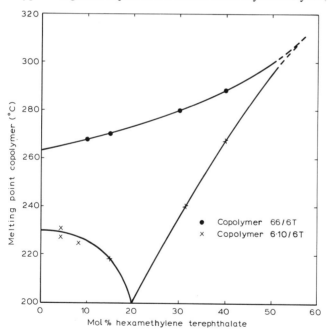

Figure 3.5 Melting point vs composition for copolymer types (courtesy *Journal of Polymer Science*)

properties depending on it, for example melting point, depend largely on composition only.

In the second category are those copolymers where the sizes of the comonomer segments differ, and replacement one by the other leads to longitudinal disorder and reduced crystallinity. In this case the melting point–composition curve (*Figure 3.5*) exhibits an eutectic of lowest melting composition corresponding to the composition of lowest crystallinity. This case is exemplified by the systems nylon 66/6 and nylon 6.10/6T.

While the isomorphous type of copolymer offers possibilities of obtaining a wide range of properties within those of the homopolymers, in that the composition corresponding to the property required can be selected, the number of useful isomorphous types is small, and the number commercially available is even smaller. Anisomorphous copolyamides, on the other hand, can exhibit properties such as high elasticity and flexibility differing so widely from those of the homopolymers that they can be commercially useful in their own right for coatings, adhesives, and the like. Copolymers of nylon 66 and nylon 6 with such properties are commercially available.

## 3.8 Cyclic oligomers

Oligomers may be defined as substances composed of molecules containing a few of one or more species of atoms or groups of atoms (mers) repetitively linked to each other. In addition to polymers and oligomers with end-groups, polyamides of all types (particularly nylon 6 and 66) contain a small proportion of cyclic oligomers without end groups which, while playing a not very significant role in the main equilibrium reactions, nevertheless modify significantly the mechanical and physical properties of the polymer.

The cyclic oligomers appear in the aqueous or alcoholic extracts of all polyamides, the main constituent being the cyclic dimer. The proportion of cyclic oligomers in equilibrium with the main polymer increases with the temperature and water content of the system. *Figure 3.6* from the experimental data of Wiloth[23] shows the relationship between the equilibrium cyclic oligomer content and the water content of the system water/caprolactam/polycaprolactam. Subsequently German workers isolated the lower molecular weight constituents of nylon 6 and nylon 66 extracts using chromatographic methods and proved the existence of cyclic oligomers containing up to nine carbon atoms in the ring. These and the dimer and trimer from nylon 11 were later synthesised.

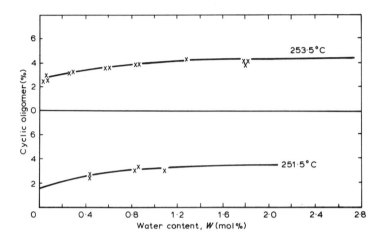

Figure 3.6  Equilibrium concentration of cyclic oligomers vs water content for nylon 6

Oligomers may be formed by cyclisation of chain ends of low molecular weight polymers or by direct cyclisation of short segments of linear polyamides. These are the likely mechanisms for the cyclic oligomers up to the nonamer. For the larger rings a more probable mechanism is through cross-amidation between parallel chains of the polymer.

The dry cyclic oligomers, particularly the caprolactam dimer, are very stable and can be heated for prolonged periods above their melting points. Polymerisation takes place in presence of water and the usual equilibrium is established. *Figure 3.7*[24] shows the relationship between the melting point of oligomers and their degree of polymerisation (DP). From the figure it can be observed that the melting point for all the cyclic oligomers lies above that of the high polymer, while the melting point curve of the linear oligomers lies below. With increasing degree of polymerisation these melting points gradually approach that of the high polymer. The extremely high melting point of the cyclic dimer, reflecting its high stability, should

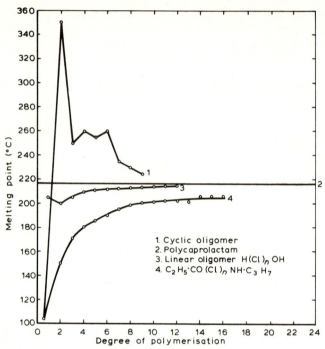

Figure 3.7  Melting point vs DP for oligomers of caprolactam (courtesy *Angewante Chemie*)

be noted. The following structures for the lower cyclic oligomers have been established:

*Cyclic dimer*

$$(CH_2)_5 \begin{array}{c} \text{—NH·CO—} \\ \text{—CONH—} \end{array} (CH_2)_5 \quad \text{(melting point 347°C)}$$

or the isomer

$$(CH_2)_4 \begin{array}{c} \text{—CONH—} \\ \text{—CONH—} \end{array} (CH_2)_6 \quad \text{(melting point 253°C)}$$

*Cyclic trimer*

$$(CH_2)_5 \begin{array}{c} \text{—NH·CO——}(CH_2)_5\text{—} \\ \text{—NHCO——}(CH_2)_5\text{—} \end{array} \text{NHCO} \quad \text{(melting point 247°C)}$$

## 3.9 Preparative methods, based on physico-chemical principles used

### 3.9.1 MELT POLYMERISATION

One understands by melt polymerisation the homogeneous single-phase condensation of mono- or bifunctional components, proceeding to a state of equilibrium at some temperature above, but not far from, the melting point of the polymer. It is necessary in this type of polymerisation to keep above the melting point in order to avoid formation of solid products that might interfere with the attainment of equilibrium.

To be effective it is necessary that both the monomer and polymer are thermally stable just above the melting point for as long as it is required for equilibrium to be established. For example, there is little danger from degradation with hydrolytic nylon 6 (melting point about 215°C), whereas during polymerisation of AH-salt to nylon 66 (melting point 265°C) the polymer is subject to appreciable degradation above the melting point and precautions such as using protective inert atmospheres must be taken to reduce this tendency.

A further requirement for effective melt polymerisation is to ensure that the components can react in a single condensed, liquid, phase. With nylon 6, for example, the monomer (caprolactam, melting point 69°C) can be introduced to the system as a melt; with nylon 66 the AH-salt starts off as an aqueous solution.

A final requirement is to prevent loss of reactants in the early stages of the polymerisation. For example, in preparing nylon 66 the hexamethylene diamine tends to volatilise, and only by working in a closed vessel is its loss prevented. This closed-system working also allows inert atmospheres to be maintained over the reactants, reducing oxidation and thermal degradation.

### 3.9.2 SOLID-STATE POLYMERISATION

If free end-groups are still available in the course of polymerisation with AB and AA BB type polymers, further condensation to higher molecular weight is possible provided the water vapour concentration over the condensed phase is below the equilibrium value for the conditions obtaining. In these conditions polymerisation can proceed in the solid state. The reaction rate is accelerated by the subdivision of the solid particles and by reducing the water vapour concentration by application of vacuum.

The technique of solid state polymerisation is often used as a second stage of manufacture of polyamides after the polymer has been brought to a certain molecular weight in the conventional type of polymerisation, and indeed it is sometimes preferred to polymerisation in the melt where removal of gaseous condensation product from a viscous melt is technically difficult.

When the polyamide is liable to thermal degradation above the melt temperature, an additional advantage accrues from second-stage polymerisation in the solid state. The velocity of polymerisation is not substantially reduced by working at temperatures below the melt, as shown by the figures in *Table 3.2* referring to nylon 66[25].

**Table 3.2** SOLID STATE POLYMERISATION RATE: NYLON 66

| Mean molecular weight, $\overline{M}$, of pre-polymer | Time (hours) to attain $\overline{M} = 15\,000$, at 216°C |
|---|---|
| 1000 | 16 |
| 2500 | 4 |
| 4000 | 2 |

Solid-state polymerisation is especially advantageous if branching or cross-linking components are incorporated in the mix. Here traces of acids such as phosphoric or sulphuric can accelerate the condensation reaction, and degrees of polymerisation of over 1000 can be attained.

### 3.9.3 LOW-TEMPERATURE POLYMERISATION

The techniques of low-temperature polymerisation as applied to polyamides are recent innovations, and were developed to enable rapid polymerisation to higher molecular weights to be attained. They are also used in the preparation of polyamides, for example those incorporating aromatic nuclei, which are unstable at melt temperatures.

The two types of low-temperature polymerisation that can be distinguished—interfacial polymerisation and solution polymerisation—are described below.

Basic chemistry and manufacture 51

### 3.9.3.1 Interfacial polymerisation

In this method polymerisation occurs at the interface between water containing a bifunctional intermediate (for example a diamine) and an inert, water-immiscible organic solvent containing the other bifunctional component (commonly a diacid chloride). A typical example is the reaction of piperazine with sebacyl chloride, which proceeds as follows:

$$HN\begin{array}{c}CH_2-CH_2\\ \\ CH_2-CH_2\end{array}NH + Cl-\overset{O}{\underset{\parallel}{C}}-(CH_2)_8-\overset{O}{\underset{\parallel}{C}}-Cl$$

(aqueous phase)    (organic solvent phase)
(+ NaOH)

$$-\left[N\begin{array}{c}CH_2-CH_2\\ \\ CH_2-CH_2\end{array}N-\overset{O}{\underset{\parallel}{C}}-(CH_2)_8-\overset{O}{\underset{\parallel}{C}}\right]_x + NaCl$$

Caustic soda is required in the aqueous phase to neutralise the acid formed in the reaction.

A summary of the essential differences between melt polymerisation and interfacial polymerisation is shown in *Table 3.3*.

**Table 3.3** COMPARISON OF POLYMERISATION METHODS

| Features for comparison | Melt polymerisation | Interfacial polymerisation |
|---|---|---|
| *Reactants* | | |
| (a) Purity | High | Medium to high |
| (b) Heat stability | Must be high | Need not be high |
| *Polymerisation* | | |
| (a) Completion time | Hours | Minutes |
| (b) Temperature | Above 200°C | Up to 40°C |
| (c) Pressure | High or vacuum | Atmospheric |
| (d) Equipment | Closed vessel and special | Open vessels of simple design |
| *Reaction products* | | |
| (a) Yield | Low to high depending on conditions | High |
| (b) By-products | Salts | Volatile materials |

## 52  Basic chemistry and manufacture

### 3.9.3.2  Solution polymerisation

This method of polymerisation is a development of the interfacial method, but in this case a non-aqueous medium (which is also a solvent or swelling agent for the formed polymer) is used. The method is particularly used for diamines of low basicity or for readily hydrolysed acid chlorides. Stirring must be rapid and maintained continuously. As here the polymer remains essentially in solution it must be recovered by precipitation and separation.

## 3.10  Molecular weight control

The properties of the final polymer obtained by the organic chemical mechanisms described in this chapter, and to some extent the type of mechanism operating in the final stages of polymerisation, depend upon the average molecular weight attained by the polymeric species.

Control of molecular weight is, therefore, essential in the technical manufacture of polyamides. Knowledge of the kinetics of the various reactions involved and equilibrium data for the final state is a prerequisite to effective control. For the commoner polyamides such data are well documented and the principles are fully described in many excellent text books on the subject.

For the polyamides we have described in this chapter control methods vary with reaction type.

### 3.10.1  STEP-REACTION POLYMERISATION

For polymerisation of groups of the AB and AA BB types the condensation will continue with the disappearance of equivalent numbers of end-groups of each type. By providing an imbalance of end-group (by, for instance, addition of small quantities of molecular weight stabilisers, which effectively render a proportion of the end-groups inactive) the linear chain grows until the end-groups in lesser amount are entirely used up.

Alternatively, with polymerisation of the AA BB type where two monomers are used, the molecular weight may be stabilised by using one monomer in small excess.

The theoretical degree of polymerisation estimated from end-group consideration is seldom attained, because loss of reactive monomer by volatilisation, the presence of reactive impurities and

*Basic chemistry and manufacture* 53

the operation of side reactions severely limit the course of the main reaction. Interchange reactions are also possible but these only affect the molecular-weight distribution and generally do not affect the average molecular weight.

Although it is also possible to reduce polymerisation rate to virtually zero by cooling from the melt, the system in this case can hardly be considered chemically stable and there will be a tendency for further condensation if the temperature is raised.

## 3.10.2 RING-SCISSION AND HYDROLYTIC POLYMERISATION

If ring-scission is due to water catalyst molecules in what we have termed hydrolytic polymerisation, all rings are opened simultaneously and the number of active centres for growth is proportional to the initial concentration of water or catalyst. The initiated chains compete for the available monomer and polymerisation proceeds with a very narrow molecular-weight distribution. To ensure such a distribution the reaction should be stopped at not too high a molecular weight. As the main reaction slows near the end-point, interchange reactions occur between the chains and the distribution of molecular weight tends to the most probable.

## 3.10.3 ANIONIC POLYMERISATION

As in other ring-scission reactions, each molecule of initiator starts a growing chain and all molecules are initiated simultaneously. In activated anionic polymerisation a fast initiation step ensures that each molecule of initiator can immediately originate an active centre for a growing chain. The absence of a conventional termination step results in a very narrow distribution of molecular weight, which can be calculated from the mole fraction of initiator in the monomer. Thus:

Degree of polymerisation (DP) = $C_m/C_i$

where

$C_m$ = initial molar concentration of monomer
$C_i$ = initial molar concentration of initiator

Generally the presence of impurities or deliberately-added modifiers terminates the polymerisation at a DP short of the theoretical.

## REFERENCES

1. Vieweg, R. and Muller, A., *Kunststoff Handbuch*, Vol. VI, *Polyamide*, 188, Carl Hanser Verlag, Munich (1966).
2. (a) Du Pont de Nemours, US Patents 2 361 717 (1940) and 2 689 839 (1951)
   (b) British Nylon Spinners, German Patent 1 158 257 (1959)
   (c) NV Onderzoekinginstitut, German Patent 1 131 011 (1956)
3. Hermans, P. H., Heikens, D. and van Velden, P. F., *J. Polymer Science*, **30**, 81 (1958)
4. Wiloth, F., *Makromoleculare*, **30**, 189 (1959)
5. Reimschüssel, H. K., *J. Polymer Sci.*, **41**, 457 (1959)
6. Du Pont de Nemours, US Patent 2 251 519 (5.8.1941)
7. Monsanto Chemical Company, US Patent 3 017 391 (1962)
8. Sebenda, J. and Králiček, J., *Collection Czech Chemical Communications*, **23**, 766 (1958)
9. Dacks, K. and Schwartz, E., *Angewante Chemie*, **74**, 540 (1962)
10. Klare, H. and co-workers, *Synthetic fibres from polyamides—technology and chemistry*, Akademie-Verlag (1962)
11. Reimschüssel, H. K., *J. Polymer Sci.*, **41**, 457 (1959)
12. Hoftyzer, P. J., Hoogschagen, J. and von Krevelen, D. W., 'Optimisation of caprolactam polymerisation', in *Chemical Reaction Engineering*, Proc. 3rd European Symp., Amsterdam, Pergamon (1965)
13. Hopff, H., Müller, A. and Wenger, F., *Die Polyamide*, Springer Verlag (1954)
14. Du Pont de Nemours, US Patents 2 628 216 and 2 628 219 (12.2.1953)
15. Franke, W. K. and Müller, K. A., *Chemie Enginieur Technologie*, **36**, 960 (1964)
16. Nesmejanow, A. and co-workers, *Chemical Technology* (USSR), **9**, 139 (1957)
17. Müller, E., *Angewante Chemie*, **64**, 233 (1952)
18. Reppe, W., Schlichtig, O., Klager, K. and Toepel, T., *Liebigs Ann. Chem.*, **560**, 1 (1948)
19. Ziegler, K. and Wilms, H., *Angewante Chemie*, **59**, 177 (1947)
20. Mark, H. F. *et al.* (eds), *Encyclopaedia of polymer science and technology*, Vol. 10, 197, Interscience (1969)
21. Morgan, P. W., *Condensation polymers*, 190–193, Interscience (1965)
22. Edgar, O. B. and Hill, R., *Polymer Science*, **8**, 1 (1952)
23. Wiloth, F., *Zeitung physikalische Chemie*, **5**, 66 (1955)
24. Zahn, H. and co-workers, *Angewante Chemie*, **68**, 229 (1956)
25. Du Pont, US Patent 3 031 433 (1958)

# 4

# Properties of polyamides

## 4.1 Chemical properties

### 4.1.1 MOLECULAR STRUCTURE

In our discussion of the basic chemistry of polyamide manufacture in Chapter 3 several types of chemical linkages were mentioned. These were: (a) linear, (b) branched, (c) cross-linked and (d) cyclic. The type of linkage present in the polymer greatly influences its molecular structure; also, in the case of polyamides, the degree of crystallinity and the size and number of spherulites depend on the symmetry and steric factors, which in turn relate to the type of linkage and ability of chains to approach each other, and allow secondary forces to come into play.

#### 4.1.1.1 Molecular weight of polyamides and relationship to properties

Polyamides with a simple linear structure are those in which the repeat units, or mers, are connected through their end functional groups to form a long chain. The length of the chain is specified by the number of repeat units it contains; this is usually called the 'degree of polymerisation', commonly denoted by the abbreviation DP. Usually the repeat unit is equivalent to the monomer. The molecular weight of the polymer is the product of the molecular weight of repeat unit and the degree of polymerisation.

Since the length of the polyamide chain in both step reaction and ionic processes is determined by random events, the polymerised product must contain chains of different lengths, and a molecular weight distribution can be calculated or determined experimentally for each system. Normally the experimental determination of molecular weight gives only an average value, but both the average and the distribution are important in determining polymer properties.

Several different averages can be distinguished, of which, at the present time, the two most important are the number-average molecular weight, $\overline{M}_n$, and the weight-average molecular weight, $\overline{M}_w$. $\overline{M}_n$ is measured by those methods that in effect count the number of molecules in a known mass of material—for example, end-group analysis and methods based on colligative properties such as ebulliometry, cryoscopy and osmometry. It should be pointed out that the number-average molecular weight is very sensitive to changes in the weight fractions of low molecular weight species, particularly when end-group analysis is the method used.

Measurement by osmometry largely discounts very low molecular weight material, which passes through the semi-permeable membrane used in the osmometer. On the other hand, accuracy of measurement of osmotic height is low for high molecular weights over 1 000 000, where the osmotic effect is low.

While the methods of end-group analysis and osmometry are popular for polyamides, the former particularly for those polyamides formed by step-reaction polymerisation, they are limited to polyamide species that can conveniently be dissolved in suitable solvents. For insoluble or partially soluble polyamides, or those of very high molecular weight, other methods must be used.

Weight-average molecular weight, $\overline{M}_w$, is usually determined by light-scattering or sedimentation methods. The equipment required for these methods is generally expensive and sophisticated, and in the case of sedimentation by the ultracentrifuge the time required for a determination is time-consuming—measured, in some cases, in weeks. In the light-scattering method, determination of molecular weight is dependent on the principle that the intensity of scattering of the incident light is proportional to the square of the mass of polymer species. In the ultracentrifuge method a solution containing the heterodisperse polyamide species is held in a cell, which is rotated over a long period. A final thermodynamic equilibrium is reached in which the polymer is distributed in the cell according to its molecular weight and molecular weight distribution, the force of sedimentation on the species being just balanced by its tendency to diffuse back against the concentration gradient arising from the centrifugal field.

Probably the most useful and widely used methods for determining the molecular characteristics of polymers in general, and polyamides in particular, are those that involve measurement of solution viscosity. Molecular weights found by such methods are termed viscosity-average molecular weights. The size and spatial influence of the polymer molecule is reflected in its velocity in a

## Properties of polyamides

solvent, and for linear polymers an empirical relationship between the viscosity and viscosity-average molecular weight can be established. Very precise measurements of solution viscosity can be rapidly carried out using the simplest of equipment, generally an efflux viscometer. The efflux time required for a known volume of polymer solution to pass through the capillary section of the viscometer is measured, as is the efflux time taken by the solvent alone to pass. The solvent type, polymer concentration and temperature must be specified for complete definition of the system. Confusion is sometimes caused by the many terms used for solution viscosity, and in some trade and technical literature these terms are loosely used often without specifying the important factors such as the solvent used. *Table 4.1* summarises the terms and the equations defining them.

**Table 4.1** TERMS USED IN SOLUTION VISCOSITY

| Common name | Recommended name | Symbol and derivation |
|---|---|---|
| Relative viscosity | Viscosity ratio | $\eta_r = \eta/\eta_0 \approx t/t_0$ |
| Specific viscosity | Specific viscosity | $\eta_{sp} = \eta_r - 1$ $= (\eta - \eta_0)/\eta_0$ $\approx (t - t_0)/t_0$ |
| Reduced viscosity | Viscosity number | $\eta_{red} = \eta_{sp}/c$ |
| Inherent viscosity | Logarithmic viscosity number | $\eta_{inh} = (\log \eta_r)/c$ |
| Intrinsic viscosity | Limiting viscosity number | $[\eta] = (\eta_{sp}/c_0)_{c=0}$ $= [(\log \eta_r)/c]_{c=0}$ |

where  $\eta$ = dynamic viscosity, solution
 $\eta_0$ = dynamic viscosity, solvent
 $t$ = efflux time, solution (seconds)
 $t_0$ = efflux time, solvent (seconds)
 $c$ = concentration (usually g/100 ml)

The empirical relation between viscosity and molecular weight was derived from an extension of Staudinger's early work[1]. The relationship itself

$$[\eta] = K'M^a$$

gives a viscosity-average molecular weight, $\overline{M}_v$. Both $K'$ and $a$ are functions of the solvent as well as the type of polymer.

The relationship between $\overline{M}_v$ and $\overline{M}_w$ depends on the molecular-weight distribution; usually $\overline{M}_v$ is slightly less than $\overline{M}_w$. The relationship can be found by determining the index $a$ experimentally. The weight distribution of chain lengths (molecular weights) for a typical polyamide is shown in *Figure 4.1*.

It should be noted that $\overline{M}_w$ is always greater than $\overline{M}_n$ except for a monodisperse system. The ratio $\overline{M}_w/\overline{M}_n$ is a measure of the polydispersity of the system.

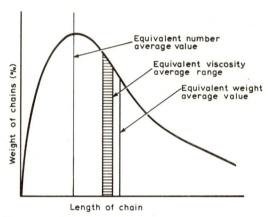

Figure 4.1  Weight-distribution of chain lengths (molecular weights) for a typical polyamide

Molecular-weight methods of polyamide characterisation are described comprehensively in a publication by Allen[2].

It is difficult to separate the effect *per se* of molecular weight and molecular-weight distribution of a polyamide on its properties, because the effect of crystallinity introduces other variables. In the virtual absence of crystallinity, i.e. in the melt, viscosity increases with number-average molecular weight. In the case of nylon 6, for example, the following relationship holds at constant temperature:

$$\log \eta = A + C \log \overline{M}_n$$

where

$\eta$ = melt viscosity, in poises

$\overline{M}_n$ = number-average molecular weight

$A, C$ = constants

The melt viscosity of a polyamide is also dependent on chain branching, cross-linking, monomer content and, of course, temperature. Molecular weight and its distribution are very important in the processing of polyamides whether this be by moulding, extrusion or

## Properties of polyamides

any other method. Ease of processing is usually favoured by lower molecular weight, but the spread is fairly critical for each process. The bottom limit for molecular weight is determined by the need to preserve mechanical properties in the final polymer.

### 4.1.2 MORPHOLOGY OF POLYAMIDES: ORDER AND DISORDER

Analysis by X-ray diffraction has shown that linear homopolyamides such as nylon 6 and nylon 66 are, in the solid state, only partially crystalline. The degree of crystallinity never reaches 100 per cent and is generally below 50 per cent. The size of single crystals is very small, seldom exceeding 200 angstroms in length, while the size of the polyamide molecule itself is of the order of 1000 Å.

In polyamides, as in other polymers, crystallinity is favoured by such features of the molecule as a high degree of regularity in space (stereo-regularity) of the functional groups, the small molecular volume of these groups, and the ability of the secondary valence forces to develop a regularly packed array of atoms.

On cooling the polymer from the melt, in which there is complete molecular disorder, the functional groups of one molecule during random motion of the segments become favourably aligned with those of a neighbouring molecule and a crystalline ordered region is formed. Other segments of the same molecular chain may lie outside the ordered region and show no crystalline characteristics. Thus single chains may pass through an ordered crystalline and a disordered amorphous region. This is the basis of the 'fringed micelle' model diagrammatically represented in *Figure 4.2*.

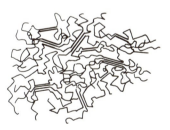

Figure 4.2  Diagrammatic representation of the fringed micelle model of order in a polyamide molecule

In the case of crystallisation from the melt in the absence of bulk flow, directional effects are absent and the crystals show random orientation. In the commonly used processes of moulding and extrusion the crystallites show anisotropy due to partial alignment of the chains along the direction of flow.

The secondary valence forces operative in crystallisation of polyamides can be largely identified with the hydrogen bond. Stereoregularity and the hydrogen bond are generally recognised to be the main structural features of polyamides that result in the high melting point of, for example, nylon 66 and nylon 6. This is illustrated diagrammatically in *Figure 4.3*.

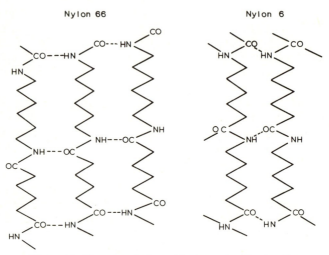

Figure 4.3  Stereo-regularity and hydrogen bonding in nylons

Hydrogen bonding is not exclusive to the crystalline fraction of the polymer. Infra-red measurements by which the bonding is detected show a very small proportion of free NH groups, and therefore extensive bonding, in polyamides cooled freely from the melt. Rapid cooling, such as can occur in injection moulding of polyamides, yields a large proportion of amorphous material and a considerable proportion of free NH groups.

While the X-ray patterns of polyamides had for long been interpreted in terms of the contribution of the crystalline and amorphous regions of the fringed micelle model, it soon became apparent from optical spectroscopy using polarised light that there exist ordered regions, much larger than the crystallites interpreted from the X-ray evidence. These regions are termed spherulites and they are recognised by their characteristic appearance in the polarising

## Properties of polyamides

microscope, where they are seen normally as circular birefringent areas showing a dark Maltese cross pattern as illustrated in *Figure 4.4*.

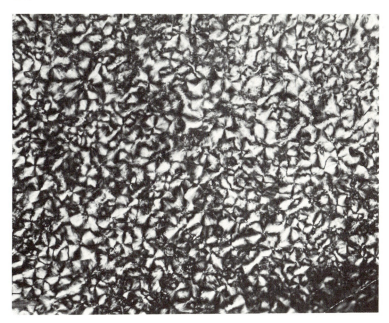

Figure 4.4    Spherulites in nylon 6

It has been shown that the spherulites in polyamides are completely crystalline while the non-spherulitic material is amorphous. The spherulites are generally accepted as primary products of crystallisation, and they commonly grow from a nucleus (which is often a foreign particle) but may also arise spontaneously. Electron-microscope studies indicate that spherulites have a lamellar structure and that crystallisation proceeds by the growth of these lamellae.

Spherulites can be classed as positive or negative according to their behaviour in polarised light. According to the type of polyamide and the growth conditions, the polyamide section viewed in the microscope can show one of a number of well defined characteristic patterns that relate to the history of the polymer. Microscopic studies can therefore be used to investigate processing faults, and the patterns observed correlate with the mechanical properties of the polymer. In the first instance, the number and size of the spherulites formed on cooling a polyamide from the melt depend on the rate of cooling and on the number of nuclei present or formed during the cooling. A small number of large spherulites are formed on slow cooling and crystallising from a small number of nuclei, while with rapid cooling and growth from a large number of nuclei the reverse is the case. Price[3] has shown that the radial growth of spherulites is a linear function of time. This is illustrated in graphs plotted from the

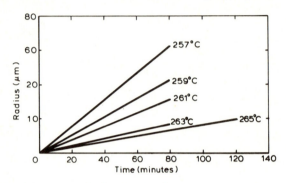

Figure 4.5  Growth rates of nylon 66 spherulites at various temperatures (courtesy *Journal of Polymer Science*)

data of Khoury[4] (*Figure 4.5*), which indicate that the spherulitic growth rate is independent of diffusion rate.

### 4.1.3  EFFECT OF CHEMICALS

#### 4.1.3.1  Permeability to gases and liquids

The permeability of thin sections of polyamides to fluid depends on the type and degree of crystallinity of the polymer. Other things being equal, permeability decreases with increasing $CH_2/CONH$ ratio in the polymer; for example, the permeability of nylon 66 to water vapour at 20°C is approximately three times greater than that of nylon 11. For a given polymer type, however, permeability decreases strongly with increase in crystallinity. Molecular orientation also has the same effect as crystallinity in decreasing permeability in polyamides.

The permeability of polyamides to other gases, in contrast to water vapour, is relatively small when strong intermolecular forces such as

## Properties of polyamides

hydrogen bonding are involved. *Table 4.2*[5] shows values of permeability of nylon 6 foil to some common gases. The permeability of polyamides to gases increases with moisture content of the polymer, due to increased chain mobility in the amorphous portion of the polymer.

**Table 4.2** PERMEABILITY OF NYLON 6 TO INORGANIC GASES (courtesy Carl Hanser Verlag)
Foil thickness = 100 µm, temperature = 30°C

| Gas | Nitrogen | Oxygen | Hydrogen | Carbon dioxide |
|---|---|---|---|---|
| Permeation rate ($cm^3/dm^2 \cdot 24h \cdot atm$) | 0.07 | 0.3 | 6.0 | 2.0 |

Polyamides are comparatively impermeable to fluid hydrocarbons but relatively permeable to alcohols. Values for other common organic solvents fall between the limits for the two types mentioned, as shown in *Table 4.3*[6]; these values should not be considered as

**Table 4.3** PERMEATION RATE OF SOLVENT VAPOUR THROUGH NYLON 6 FILM (courtesy *Kunststoffe*)
Film thickness = 1 mm, temperature = 20°C

| Solvent | Permeation rate ($g/24h \cdot m^2$) |
|---|---|
| n-Hexane | 0.420 |
| Benzine | 0.016 |
| Benzene | 0.053 |
| Toluene | 0.130 |
| Sulphur dioxide | 0.410 |
| Chloroform | 2.800 |
| Carbon tetrachloride | 0.030 |
| Methanol | 4.500 |
| Ethanol | 5.100 |
| Isopropanol | 2.100 |
| Diethylether | 0.840 |
| Isopropylether | 0.072 |
| Dioxan | 0.038 |
| Acetone | 0.230 |
| Methyl ethyl ketone | 0.022 |
| Acetic acid | 0.190 |
| Ammonia | 1.400 |
| Diethylamine | 0.001 |
| Pyridine | 0.005 |

absolute. It is expected that linear polyamides, other than nylon 6, would show the same ranking order of permeation rate to solvents as in this table.

Since it has been shown that permeability is not linear with thickness of section, the data shown in the tables should be used more as a guide to permeability behaviour rather than for design purposes. Permeability data is sparse in the literature on polyamides and raw material manufacturers' data sheets provide as good a source as any, particularly as the permeation rates for commercial fluids are generally quoted.

### 4.1.3.2 Compatibility with chemicals

*(a) Factors influencing behaviour*

An understanding of the basic factors influencing the interaction of chemicals with polyamides aids in the prediction of the compatibility behaviour in service. Manufacturers of nylon usually issue their own compatibility tables for commercial chemicals in contact with the common types of nylons. In the writer's experience these publications serve only as approximate guides. The less common chemicals are seldom listed.

In the organic field the molecular structure of a chemical and the functional groups it contains are the most important factors determining its compatibility with polyamides. In the inorganic field the behaviour of acids and their aqueous solutions depends upon the mobility of the hydrogen ion and its interaction with the amide (—CONH—) group. Oxidising acids such as nitric acid can in addition attack at points along the methylene chain of a polyamide and bring about chain scission. Inorganic salts are generally compatible but several can attack the surface of polyamides if this is stressed. Chemical action on polyamides increases with temperature, as would be expected. The effect can be solely physicochemical (usually dependent on diffusion of the fluid into the polymer), purely chemical (where chemical groupings in the fluid react with polymer functional groups), or a combination of the two.

The crystallinity of the polyamide is an important factor determining the rate of diffusion or chemical interaction between fluid and polymer. Generally diffusion rates are smaller and chemical action is less the more crystalline the polyamide.

The $CH_2/CONH$ ratio in the polyamide affects the susceptibility to attack by chemicals; the greater this ratio the less prone is the polyamide to be attacked (see *Table 4.4*). The interchain forces between neighbouring methylene groups are diminished, however, by hydrocarbon type solvents; this effect is more noticeable with nylon 11 and 12, where swelling rather than solution tends to occur.

Table 4.4  SOLUBILITY OF POLYAMIDES IN FORMIC ACID: EFFECT OF $CH_2$/CONH RATIO

| Polyamide | $CH_2$/CONH ratio | Lowest concentration of formic acid to effect solution |
|---|---|---|
| 6 | 5 | 70 |
| 66 | 5 | 80 |
| 6.10 | 7 | 90 |
| 11 | 10 | Insoluble |
| 12 | 11 | Insoluble |

The detailed effects of the above-mentioned factors are discussed in the following sections.

*(b)  Mineral acids, bases, salts, and pH effect*

The more common mineral acids and their aqueous solutions swell, dissolve or hydrolyse polyamides according to the concentration. In 1 N solution, hydrochloric and nitric acids are more effective than sulphuric acid since at that concentration the hydrogen ion concentration is smaller in the latter due to the effect of first stage dissociation only (i.e. $H_2SO_4 \rightleftharpoons H^+ + HSO_4^-$). Nitric acid is more effective than hydrochloric acid at medium to high concentration since the former has oxidation properties that lead to additional breakdown and solvent action. For this reason nylons 11 and 12 can be dissolved in nitric acid at concentrations where sulphuric or hydrochloric acid would be ineffective.

The dilute acid solution viscosity of polyamides may be used for characterising the polymer with regard to molecular weight, but with increasing concentration of the acid and increase in temperature the hydrolysis begins to break down the polymer chain. The rate of this process may be measured by observing the decrease in solution viscosity.

Solvent effects in aqueous acids decrease with approach of the pH value to the neutral point. In continuous service at room temperature immersion of polyamides in solutions more acid than pH 4 is not recommended. On the alkaline side of the neutral point polyamides are remarkably stable, and a high concentration or high temperature is required before chemical attack is observed. This is due to the absence of the hydrated hydrogen ion, $H_3O^+$.

In dilute mineral acids at concentrations lower than those required for solution and hydrolysis, polyamides are subject to a form of corrosion that shows up as a network of fine cracks or fissures of various depths on the surface of parts immersed in the fluids. In 2–5 per cent hydrochloric acid solution at room temperature these cracks can appear in nylon 6 and 66 after a few months.

In neutral solutions of some inorganic salts a type of stress corrosion can occur, appearing as a maze of fine cracks on the surface of the part exposed to the solution. The part itself must be under stress, usually in flexure. This form of stress corrosion may be eliminated by moisture-conditioning the part before introduction to service. Thin-wall nylon tubing is particularly prone to stress-corrosion, one of the offending salts being zinc chloride. At one time automobile manufacturers banned the use of nylon 6 tubing for auto fluid lines because of stress corrosion susceptibility, but compounds can now be formulated specially to overcome this trouble. One of the standard test methods used to measure stress corrosion susceptibility of plastics, including polyamides, is detailed in ASTM D-1693. Here a standard test specimen is subjected to a predetermined stress in contact with the corrosion medium, and the surface periodically examined for cracks.

*(c) Influence of diffusion*

The rate of diffusion of solvents into thick-section polyamide depends on the solvent type, concentration and temperature and, in the case of aqueous acids, the pH value. *Table 4.5* lists the diffusion coefficients of acid solvents for nylon 6. The values in the table were obtained using nylon test specimens incorporating uniformly dispersed bromophenol blue indicator. The permeation of the specimen by the hydrogen ion of the acid as a function of time was measured by cutting sections of the specimen at regular time intervals and observing the rate of advancement of the colour fringe, indicating acidity, towards the centre of the specimen.

It should be noted that the rate of hydrolytic attack on thick-

## Properties of polyamides

**Table 4.5** DIFFUSION COEFFICIENT OF SOLVENT ACIDS IN NYLON 6 AT 25°C

| Aqueous acid (1 N) | Diffusion coefficient, D ($\times 10^8$ cm$^2$/s) |
|---|---|
| Citric | 0.132 |
| Phosphoric | 0.386 |
| Acetic | 1.409 |
| Hydrobromic | 1.530 |
| Sulphuric | 1.865 |
| Hydrochloric | 1.954 |
| Nitric | 1.981 |
| Formic | 2.416 |
| Sodium hydroxide (for comparison) | 0.011 |

section polyamide is determined not only by the rate of hydrolysis, which is generally a first-order reaction, but also by the rate of diffusion. In cases where the diffusion rate exceeds the hydrolysis rate, the kinetics of the process are governed by the chemical reaction. Where the diffusion rate is less than that of the hydrolysis the process is diffusion-controlled.

An Arrhenius plot of heterogeneous hydrolysis rate against reciprocal of absolute temperature can sometimes reveal a change in the process with temperature. Thus a change in slope of the Arrhenius plot for heterogeneous hydrolysis of nylon 6 in 0.1 N sulphuric acid occurs at 71°C and has been put down to the greater activation energy required for hydrolysis of the crystalline fraction of the nylon, which predominates over that of the amorphous fraction above that temperature.

### (d) Oxidising agents

It has already been mentioned that nitric acid in concentrated aqueous solution, in addition to possessing hydrogen-ion activity, has potential as an oxidising agent. Using this acid, attack at points along the methylene chain of polyamides is therefore possible.

Other common oxidants are the halogen gases (chlorine and bromine), their aqueous solutions (chlorine and bromine water), iodine–potassium iodide solutions, aqueous potassium permanganate and hydrogen peroxide.

### (e) Organic solvents

Organic acids known to be good solvents for the polyamides of lower $CH_2/CONH$ ratio are formic acid and the chloracetic acids. Formic acid has found particular favour as a room-temperature solvent for preparation of solutions intended for solution-viscosity measurements that may be used to determine molecular weights. Since the hydrolytic activity of formic-acid solutions is very low at normal ambient temperatures, the solutions are quite stable over a long period.

Hydroxy-aromatic compounds are another group well known as solvents for polyamides. This group includes phenol, ortho- and para-cresols and resorcinol. The solvent action is often enhanced by addition to the aqueous solution of methanol or ethanol.

The lower molecular weight fractions of some polyamides are soluble in methanol and ethanol. The higher boiling alcohols, glycols, lactams and lactones used in the temperature range at 150–200°C are solvents for medium molecular weight polyamides, and the solutions are used in the manufacture of fine polyamide powders which are obtained by subsequent cooling of, or water addition to, the solutions. Benzyl alcohol and phenyl ethyl alcohol have also been used as high-temperature solvents for polyamides.

Chlorinated organic compounds are among the best swelling agents for polyamides. Trichloracetic acid has been mentioned, but methylene chloride, chloroform and tetrachlorethylene cause swelling or even solution under certain conditions. Carbon tetrachloride, on the other hand, has no swelling or solvent actions against polyamides. The important feature here is the molecular symmetry of the tetrachloride. The other chlorocompounds mentioned are asymmetric and have strong dipole moments capable of interaction at the sites of hydrogen bonds along the molecular chain of the polyamide. It is of interest to note also that tetrachlorethylene, which retains some aliphatic character, swells nylon 11 more than nylon 6 or 66.

### (f) Swelling and plasticisation

Swelling arises from the interactions of the dipole and hydrogen bonding forces in the swelling agent or solvent with those of the polyamide. Plasticisation on the other hand arises when the plasticiser and polyamide have similar intermolecular forces and

## Properties of polyamides

structures, when mutual miscibility is possible.

Semi-crystalline linear polyamides are not easily plasticised, but a number of good plasticisers for copolyamides exist. These include compounds containing hydroxyl or amide groups such as dihydroxy-diphenyl, toluene sulphonic acid amide, and ε-caprolactam. As with other polymers, the plasticised polyamide retains the outward appearance of the original polymer, the main effect of the additive being to reduce glass transition temperature, elastic modulus and hardness. At the same time the temperature resistance and maximum service temperature may be reduced.

### (g) Copolyamides, branched and N-substituted polyamides

The lower degree of crystallinity possessed by these polyamides admits of a much wider range of solvents and greater solubility. Alcohols and chlorinated hydrocarbons, for instance, may be used at room temperature to obtain solutions of relatively high concentration, many of which have technical importance—for example for polyamide-based adhesives and coatings.

### (h) Compatibility table

*Table 4.6* summarises the behaviour, in contact with nylons, of many of the chemicals discussed. Only those chemicals that are known to be aggressive to polyamides are listed, and these are grouped together by type. Four nylon types, representative of the range commercially available, have been chosen to illustrate change in compatibility behaviour with $CH_2/CONH$ ratio in the polyamide.

### 4.1.4 EFFECT OF ENVIRONMENT

Although polyamides are known to be among the most environmentally resistant types of thermoplastics they are subject to degradation, like most other thermoplastics, during manufacture and in service. In the former case thermal degradation is operative, to some extent, during those steps of the manufacturing process that occur after polymerisation. In the latter case degradation is commonly due to slow oxidation, usually influenced by exposure to short-wavelength radiation in the visible or UV region of the

## Properties of polyamides

**Table 4.6** MATERIALS AGGRESSIVE TO NYLONS

| Group | Nylon 6 | Nylon 66 | Nylon 6.10 | Nylon 11 |
|---|---|---|---|---|
| ELEMENTS | | | | |
| Chlorine | | | | |
| (a) Gaseous | – | – | – | – |
| (b) Aqueous | – | – | – | – |
| Fluorine | – | – | – | – |
| Iodine (alcoholic solution) | – | – | – | – |
| INORGANIC COMPOUNDS | | | | |
| *Acids* | | | | |
| Chromic acid (aqueous 10%) | – | – | – | – |
| Hydrofluoric acid (aqueous 40%) | – | – | – | – |
| Phosphoric acid (aqueous 10%) | – | – | – | × |
| Nitric acid (aqueous 2%) | – | – | – | – |
| Hydrochloric acid (aqueous 2%) | – | – | – | × |
| Sulphuric acid (aqueous 2%) | – | – | – | – |
| *Salts* | | | | |
| Ferric chloride | – | – | – | × |
| Mercuric chloride (aqueous 6%) | – | – | – | × |
| Thionyl chloride | – | – | – | × |
| Calcium chloride (20% alcohol) | – | – | – | – |
| *Oxidising agents* | | | | |
| Hydrogen peroxide (30%) | – | – | – | – |
| Potassium permanganate (1%) | – | – | – | – |
| ORGANIC COMPOUNDS | | | | |
| *Acids* | | | | |
| Formic acid (aqueous 40%) | – | – | – | – |
| Chlorsulphuric acid | – | – | – | – |
| Acetic acid (aqueous 40%) | – | – | – | × |
| Trichloracetic acid | – | – | – | – |
| *Aromatic hydroxy compounds* | | | | |
| Phenol (molten or aqueous) | – | – | – | – |
| Resorcinol | – | – | – | – |
| Cresol | – | – | – | – |

Key to symbols: – soluble; × partially soluble

spectrum. The presence of water vapour usually has a marked effect. The combined degradation process is often termed weathering.

The mechanisms of thermal degradation of polyamides in both presence and absence of oxygen, and oxidative degradation associated with exposure to short-wavelength radiation and/or water vapour, have been fairly well established for nylon 6 and 66, although a number of anomalous features remain. There is less published work on the breakdown mechanisms of other polyamides

## Properties of polyamides

but it is generally recognised that mechanisms similar to those proposed for nylon 6 and 66 apply to most linear polyamides. The mechanisms currently assumed for these degradation processes are briefly described in the following sections.

### 4.1.4.1 Thermal degradation

*(a) Degradation in absence of oxygen*

Knowledge of the thermal degradation of polyamides in absence of oxygen or under an inert atmosphere is important in the processing of polyamides, particularly in melt spinning of fibres, since the quality of the fibre and derived yarns is greatly affected by uncontrolled degradation. In extrusion and injection-moulding also, the properties of the product are adversely affected by degradation.

In inert atmospheres, the main feature of the degradation process above the melting point of the polymer is the evolution of water and carbon dioxide (usually associated with small quantities of ammonia) and, in the case of nylon 66, still smaller quantities of cyclopentanone. On prolonged heating the residual polyamide becomes insoluble in formic acid and is considered to have 'gelled' by cross-linking. In the case of nylon 66, degradation is accompanied by a decrease in carboxyl end groups.

The amounts and proportions of the degradation products, and to some extent of the products themselves, depend upon the degradation conditions, particularly the temperature. For example, in degradation investigations carried out by Ackhammer *et al*[7] and, separately, by Strauss and Wall[8] using nylon 6, 66 and copolymers, no ammonia was found in the gaseous decomposition products. On the other hand, using milder degradation conditions more likely to be encountered in processing, Kamerbeek *et al*[9] found that all the gaseous products mentioned above were evolved. The degradation mechanisms proposed by the latter workers account for the evolution of products observed and also for the phenomenon of gelation of the polymer. In their view the degradation results from primary and secondary reactions described by the following equations.

Scission occurs at the $-NH \cdot CH_2-$ group to yield one fragment containing a carbonamide end and another fragment of unsaturated hydrocarbon.

$$-CO \cdot NH \cdot (CH_2)_4 - CH_2 \cdot CO \cdot NH - CH_2(CH_2)_4 CO \cdot NH-$$
$$\downarrow \qquad (4.1)$$
$$-CO \cdot NH \cdot CH_2(CH_2)_4 \cdot CO \cdot NH_2 + CH_2 : CH(CH_2)_3 \cdot CO \cdot NH-$$

The amide group can split off water forming a nitrile:

$$-CO \cdot NH \cdot CH_2(CH_2)_4 \cdot CO \cdot NH_2 \rightarrow CO \cdot NH \cdot CH_2(CH_2)_4 \cdot C:N + H_2O \qquad (4.2)$$

Secondary reactions occur by reaction of the water produced in equation (4.2). For instance amide groups along the chain may be hydrolysed to give free amino and carboxyl end-groups:

$$R \cdot NH \cdot CO \cdot R' + H_2O \rightarrow R \cdot NH_2 + R' \cdot COOH \qquad (4.3)$$

End-groups generated as shown in equation (4.3), and of course end groups originally available, can interact to produce gaseous products and groups capable of further interaction to give branched structures. This scheme is shown below.

$$\left. \begin{array}{l} -CO \cdot NH(CH_2)_5 \cdot COOH + COOH(CH_2)_5 \cdot NH \cdot CO- \\ \downarrow \\ -CO \cdot NH \cdot (CH_2)_5 \cdot CO(CH_2)_5 \cdot NH \cdot CO- + \underline{CO_2} + H_2O \end{array} \right\} \qquad (4.4)$$

$$\overset{|}{C}=O + NH_2- \rightarrow \overset{|}{C}=N- + H_2O \qquad (4.5)$$

$$\left. \begin{array}{l} -NH \cdot CO \cdot (CH_2)_5 \cdot NH_2 + NH_2(CH_2)_5 \cdot CO \cdot NH- \\ \downarrow \\ -NH \cdot CO(CH_2)_5 \cdot NH(CH_2)_5 \cdot CO \cdot NH- + \underline{NH_3} \end{array} \right\} \qquad (4.6)$$

$$\overset{|}{N}-H + COOH \rightarrow \overset{|}{N}-\overset{|}{\underset{O}{\overset{\|}{C}}}- + H_2O \qquad (4.7)$$

While equations (4.4) and (4.6) can explain the production of $CO_2$ and ammonia, confirmation of the gelation reaction necessitates identification of the appropriate grouping in the gelled material by analysis. Kamerbeek, using paper chromatography in studying nylon 66 degradation, confirmed reaction (4.6) but was unable to find the grouping derived from equation (4.4). On the other hand from an analysis of gelled nylon 6 it seemed more likely that equation (4.4) was operative.

## Properties of polyamides

The production of cyclopentanone in the degradation reaction can be explained in the case of nylon 66 by invoking a decarboxylation reaction of adipic acid end-groups according to the following:

$$R \cdot NH \cdot CO(CH_2)_4 \cdot COOH \rightarrow R \cdot NH_2 + CO_2 + \text{cyclopentanone} \quad (4.8)$$

The above brief description of the degradation mechanism hardly does justice to the exhaustive investigation carried out by Kamerbeek and co-workers, and indeed to very many earlier workers in the field of thermal degradation. The subject is in fact still controversial.

### (b) Degradation in dry and damp air

Many mechanisms have been proposed to explain the thermal degradation of polyamides in presence of oxygen and it would be unrewarding to attempt to summarise them. From the point of view of end use and application in environments commonly encountered (that is, in presence of oxygen and water vapour) a useful study has been made by Harding and MacNulty[10], in the course of which protective treatments to reduce degradation were also investigated. The experiments were carried out in the dark to exclude photochemical effects. Taking as the criterion of embrittlement a tensile-strength loss of 80 per cent on the untreated nylon, it was found that nylon 66 embrittled with rise in temperature, as shown in *Table 4.7*.

**Table 4.7** EMBRITTLEMENT OF NYLON 66

| State | Temperature (°C) | Time to embrittle |
|-------|------------------|-------------------|
| Dry   | 70               | 2 years           |
| Wet   | 70               | 8 weeks           |
| Wet   | 90               | 4 weeks           |
| Dry   | 100              | 7 days            |
| Wet   | 100*             | 5 weeks           |
| Dry   | 150              | <24 hours         |
| Dry   | 200              | <6 hours          |
| Dry   | 250              | <2 hours          |

* Nylon totally immersed in water. Below 100°C wet conditions refer to the specimen exposed to air saturated with water vapour.

The effect on the amount of observed degradation of impregnating the specimens with protective agents was also studied. Of eight aromatic amines used, diphenylamine was most effective in preventing embrittlement, the life to embrittlement at 150°C of specimens treated with this reagent being 336 hours compared with less than 24 hours for an untreated control.

A drop in molecular weight was found with time of exposure to treatment, but this effect was limited to the surface layer of the specimen for both wet and dry conditions. In the case of treatment in dry conditions, molecular weight of the bulk of nylon increased, but for wet conditions decreased. These effects were assigned respectively to further condensation in the nylon and degradation by water, not to the normal hydrolysis at the amide bond since no increase in end-groups was observed. No satisfactory explanation was offered for this water-degradation reaction. As regards embrittlement at the surface this was shown not to be due to an increase in the crystallinity of the nylon, as might be expected, but was assigned to oxidation leading to chain-scission in the dry treatment and combined oxidation and hydrolysis with chain-scission in the wet treatment. In fact all three processes (oxidation, hydrolysis and condensation) most probably operate simultaneously in this type of degradation, the rate-determining step or the resultant effect depending on conditions of treatment. The effect of the protective treatment already mentioned is to minimise the effect of oxygen attack by preventing the chain reactions that operate in these cases. This reagent seems to have limited effect on the breakdown of nylon by hydrolysis.

It is very probable that the mechanism of the oxidation processes involved in the dry oxidation of nylons follows the classical scheme now accepted for high polymers in general, i.e. a three-step process as follows:

*Initiation*      Production of radicals $R^\cdot$

*Propagation*    $R^\cdot + O_2 \to ROO^\cdot$
                    $ROO^\cdot + RH \to ROOH + R^\cdot$

*Termination*    $\left. \begin{array}{l} R^\cdot + R^\cdot \\ ROO^\cdot + ROO^\cdot \\ R^\cdot + ROO^\cdot \end{array} \right\} \to$ Inactive products

where RH is the polyamide, $R^\cdot$ is the free radical, $ROO^\cdot$ the hydroperoxy radical and ROOH the corresponding hydroperoxide.

Degradation by the above mechanism would account for the chain scission and reduction in molecular weight observed in the oxidised layer of the nylon.

## 4.1.4.2 Light Degradation

Polyamides can be degraded by the radiant energy of visible light but more effectively by ultraviolet radiation. Much research has been carried out on the UV irradiation of polyamides in absence of oxygen, e.g. in vacuum, but more important from the practical viewpoint is a knowledge of the degradation process initiated by light in the presence of oxygen.

Basically, if a polyamide absorbs UV light, photodissociation occurs with the production of free radicals. One such scheme was proposed by Moore[11] in investigations of the photodegradation of nylon 66 under nitrogen and in air. He suggested a photolysis reaction occurring, independent of oxygen, initiated by wavelengths above 3000 Å. For example:

$$-CH_2 \cdot CH_2 \cdot CH_2 \cdot \overset{O}{\overset{\|}{C}} NHCH_2 - \xrightarrow{h\nu} (CH_2)_3 \cdot \overset{O}{\overset{\|}{C}} + NHCH_2 -$$

For wavelengths greater than 3000 Å Moore proposed that photoxidation occurred with attack at the carbon atom adjacent to the nitrogen atom. An alkoxy radical was formed which subsequently decomposed.

A peroxide mechanism was proposed by Kroes[12] for the photoxidation of nylon 6 yarn. Initiation involved an oxygen-independent scission of the C—N bond.

The photolysis and photoxidative processes may be schematically represented thus:

*Photolysis* $\qquad R \cdot COR' \xrightarrow{h\nu} RCO^{\cdot} + R^{\cdot}$

*Photoxidation* $\qquad R^{\cdot} + O_2 + RH \rightarrow ROOH + R^{\cdot}$

with the peroxide capable of dissociation to free radicals. Degradation can then proceed in accordance with the three-step process of initiation, propagation and termination described in the previous section.

## 4.1.4.3 Accelerated testing

The performance of polyamides in respect of their resistance to the combined degrading effect of UV radiation, oxygen and moisture is practically of great importance. Unfortunately the service testing in various climates required to obtain data is time-consuming, and the bases of programmes to obtain such data must be generally agreed to allow valid comparison of the results from different investigators.

In the UK it is believed that the Rubber and Plastics Research Association has embarked on such a programme, and the Ministry of Defence has been engaged for a long time in climatic testing of plastics materials. Results, if available, have not been published.

Equipment is, however, available for accelerated tests designed to simulate normal weathering conditions. Although results from these tests do not always agree with those from longer-term testing in natural environments, they give a good approximation and the method has the advantage that standard conditions reproducible in any laboratory are used. Test results obtained by different laboratories can therefore be compared.

The Fade-O-Meter, for example, is one such piece of equipment on which exposure of specimens for one hour is claimed to correspond to one average day's outdoor exposure on the 50th parallel.

Another, the Xenotest, uses a 1500-watt high-pressure xenon arc lamp. Infra-red radiation is filtered out and a fan cools the specimen to between 30°C and 35°C. Humidity is controlled within the enclosure housing the test specimens.

A third, the Weather-O-Meter, uses a carbon arc as radiation source, with complete or partial wetting conditions either in presence or absence of radiation. The conditions are controllable on a cyclic basis, together with automatic humidity control. An ASTM test method, D 1499, details test conditions that can be achieved using the Weather-O-Meter.

Chottiner and Bowden[13] of Westinghouse Electric Corporation tested over two dozen types of plastics on a 2500-hour long exposure test in the Weather-O-Meter. The plastics types included an unfilled nylon 66, an $MoS_2$-filled nylon 66, eighteen other thermoplastics, and nine thermosets. Observation of specimens was made at 300-hour intervals to observe crazing or cracking, chalking and fading when a rating chosen from five degrees of severity was assessed. Results showed the crazing performance of nylon was superior to nine of the other thermoplastics, and the chalking performance better than twelve. Incorporation of molybdenum disulphide into the nylon reduced both time to craze and time to chalk to 300 hours and 600 hours respectively. It has been said that

## Properties of polyamides

300 hours in the Weather-O-Meter is equivalent to one year's outdoor exposure in the central portion of the north temperate zone, and in this instance at least it should be possible to confirm the correlation. The authors gave no other details of the compositions of the nylons used, e.g. whether or not they were stabilised.

Each type of accelerated test has special merits, but of those mentioned above the Xenotest is considered to give a fair approximation to natural sunlight. The xenon lamp output, however, tends to decrease with age.

### 4.1.4.4 Stabilisers and their action

The patent literature is replete with claims for materials that will stabilise polyamides against the effects of heat, UV radiation, visible light and hydrolysis. Since the main function of these agents is to reduce the initiation or propagation of oxidation, thereby retarding degradation, they are termed anti-oxidants.

Anti-oxidants capable of suppressing initiation may be UV absorbers, of which carbon black is typical, or peroxide deactivactors, of which organic phosphites are typical. Probably, however, the best known anti-oxidants for polyamides are the propagation suppressors, of which the secondary aromatic amines and the hindered phenols are most widely used.

Many synergistic anti-oxidant systems for polyamides have been patented for which it is claimed that the total stabilising effect of the additives taken together is greater than the sum of the stabilising actions of the separate components.

### 4.1.4.5 Effect of fillers on degradation

Apart from carbon black and titanium dioxide, little study has been carried out on the effect of fillers on the degradation of polyamides.

As with polyolefines, certain forms of carbon black have been shown to stabilise polyamides against heat, light and general weathering. The same pigment is often used to improve the effect of the more usual stabilisers in a polyamide.

Titanium dioxide pigment has long been used as a delustrant for nylon yarn. The modification known as rutile is also an effective light stabiliser, while the anatase modification is not effective unless treated with specific agents such as manganese compounds.

Stabilised anatase pigment is preferred to rutile for delustring fibres, since the latter form has abrasive properties unsuitable in processing.

#### 4.1.4.6 Degradation from high-energy radiation

In this context high-energy radiation is considered to be that due to electromagnetic radiation of the X-ray or shorter wavelength of the spectrum, or to electrons, protons, neutrons or other high-energy particles. Radiation of this energy level causes ionisation in the irradiated material capable of producing profound changes in structure.

It has been shown conclusively by Charlesby[14] and others that nylon 66 is cross-linked by high-energy ionising radiation. Sisman and Bopp[15] found that nylon slowly loses its crystallinity and becomes clearer but darker in colour with increasing radiation dose in the Mrad range. The same workers also found that pile-irradiated nylon increased its tensile strength and modulus but also became less ductile and more brittle. A two-fold increase in tensile modulus was found for an irradiation dose of 800 Mrad.

The irradiation behaviour of polyamides is very similar to that of the polyethylenes.

#### 4.1.4.7 Biological degradation

Polyamides have been shown to be quite resistant to attack by insects—including termites, recognised to be the most aggressive type.

### 4.2 Mechanical properties of dry polyamides

The mechanical properties quoted in quality-control specifications for polyamides as raw materials, and also in many raw-material suppliers' technical data sheets, are of limited use to engineers and designers. However, abundant design data are available from the larger raw-material suppliers, relating to the effect on properties of variables such as temperature, time, moisture content and rate of loading, and including information on wear, bearing performance and environmental effects. From the designer's point of view a distinction must be made between short-term and long-term properties. Presentation of data relating to the latter is a relatively recent innovation. Mechanically, for short-term applications of load and at low stresses, polyamides exhibit a well-defined modulus. For longer-term load applications and higher stresses, creep phenomena are exhibited. At still higher stresses and at uniform strain rate, as in short-term tensile testing, polyamides generally show a well-defined yield point in the stress–strain curve and subsequently cold flow up to the point of fracture, although the yield point may be immediately followed by fracture without cold flow as occurs with very high modulus fibre-reinforced polyamides. The mechanical behaviour of engineering polyamides is therefore time-dependent. With increase in temperature, the mechanical modulus and the yield point decrease in the short term and creep increases in the long term. The mechanical behaviour of polyamides is therefore also temperature-dependent.

The following sections deal primarily with mechanical properties of dry polyamides, which is the condition when freshly moulded or extruded, or polyamides in equilibrium with standard atmospheres. The susceptibility of most polyamides to moisture uptake in certain environments can lead to a substantial change in their mechanical properties. This subject is discussed in subsequent sections.

## 4.2.1 SHORT-TERM PROPERTIES

Most mechanical-property data quoted by manufacturers and suppliers of polyamides refer to measurements taken over a comparatively short period using standard temperatures and strain rates. In this type of testing the load is normally applied at a constant rate of strain, and the method is generally used for laboratory tensile, compressive, flexural and shear specification testing.

### 4.2.1.1 Tensile properties

In tensile testing strain rates in the range 1 mm/minute to 500 mm/minute are commonly used. Within this range appreciable differences in the stress–strain relationships show up with increasing strain rate, the higher rates indicating greater moduli and yield stresses. Testing speeds should therefore always be quoted in specification and control testing, and account must be taken of this factor when comparing test results. Most quality-control testing standards specify a number of testing speeds from which a selection can be made.

The increase of yield stress with strain rate is associated with an increase of the proportional (elastic) range where stress is proportional to strain, and, usually, with a decrease in the elongation at break and a decrease in the non-linear elastic range before yield. Subjection of the polyamides to sufficiently high strain rates, such as encountered in impact loading, eliminates the yield zone and brittle fracture results. Decrease in temperature has a similar effect to increase in strain rate on modifying the shape of the stress–strain curves of polyamides. With increase in temperature the proportional elastic zone decreases and the yield stress is reduced. The effect of temperature on the tensile stress–strain curve of dry nylon 66 is shown in *Figure 4.6*.

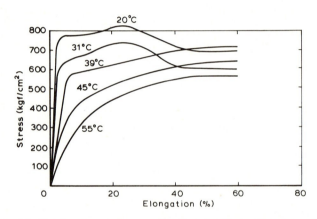

Figure 4.6 Tensile stress–strain curves for dry nylon 66 at various temperatures

As with strain rates the test temperature must always be quoted in specifying or comparing short-term tensile performance of polyamides.

### 4.2.1.2 Compressive properties

In designing polyamide components, information on the behaviour of the material under compression may be as important as knowledge of the tensile properties.

## Properties of polyamides

At low strains the moduli of elasticity of nylons in compression and tension are approximately equal. At high strains the compressive stress is larger than the corresponding tensile stress, indicating that the compressive yield stress is the greater. The difficulty of locating the yield point on the stress–strain curve has led to the common practice of defining the 0.1 or 1.0 per cent offset yield stress as indicative of the behaviour of a particular nylon in compression. *Figure 4.7*[16] shows short-term stress–strain curves for a typical nylon 66 in tension and compression. As with deformation of polyamides in tensile testing, an increase in strain rate in compressive testing results in an increase in compressive modulus and yield stress.

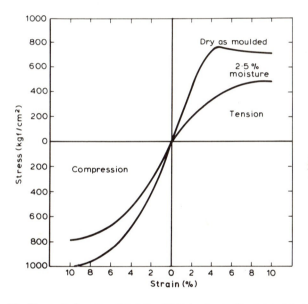

Figure 4.7  Stress–strain curves for nylon 66 in tension and compression (courtesy Du Pont de Nemours)

Increase in the testing temperature shows up in a decrease in modulus and yield stress. Strain rate and test temperature must, therefore, always be quoted when compression test data are compared.

### 4.2.1.3 Flexural properties

The determination of short-term flexural strength and modulus of polyamides is most conveniently and accurately carried out using one of a variety of the standard methods such as those described in ASTM D 790 and DIN 53452. Using the latter method the flexural modulus and yield-stress characteristic of the polyamide under the standard conditions of the test are measured. The flexural mode of deformation has the advantage of allowing accurate measurement of modulus at low strains. As with tensile and compressive tests flexural modulus and yield stress both decrease progressively as the test temperature is raised.

### 4.2.1.4 Hardness

Although the term hardness is sometimes used to denote scratch resistance or rebound resilience, the definition is restricted in the present context to describe resistance to indentation, i.e. the response of the material to a compressive deforming load applied in a particular way. In contrast to the methods used for short-term tensile, compressive and flexural tests, indentation hardness tests are generally carried out under conditions of constant load; also these tests measure the properties of the material at or near the surface only, and not throughout the bulk of the specimen. Usually the load is applied normal to the surface through a ball or needle indentor. Penetration or compression of the surface continues until stress is raised beyond the yield point of the material. The tensile yield stress can in fact be used to derive approximate values of hardness. Account must be taken of the temperature and moisture content of the material tested. The methods used for hardness measurement are described in fuller detail in a recent PRI Monograph[17], but it should be appreciated that using any method values can only be compared if the condition of applying and measuring load and the time of application (and recovery, if applicable) are all standardised. This has led to the adoption of standard testing instruments that are widely known and internationally accepted, the two most popular for polyamides being the Rockwell Hardness Tester (normally used on the R or M scale) and the Shore Durometer Type D indentation tester.

These two instruments work on slightly different principles, but both are used in specifying materials and also in quality control of

## Properties of polyamides

fabricated parts. Since the Shore Durometer can be made portable it may be used locally on complete parts at the site of fabrication. The Rockwell and Shore scales are arbitrary, the former being the more discriminating. *Table 4.8* shows typical values for the Rockwell and Durometer hardness of some common polyamides.

**Table 4.8** TYPICAL HARDNESS VALUES FOR COMMERCIAL NYLONS AT 20°C; DRY AS MOULDED/EXTRUDED

| Type | Rockwell R | Durometer D |
|---|---|---|
| Nylon 66 | 115 | 89 |
| Hydrolytic Nylon 6 | 112 | 83 |
| Cast Nylon 6 | 117 | 90 |
| Nylon 6.10 | 110 | 82 |
| Nylon 11 | 100 | 81 |
| Nylon 12 | 100 | 81 |

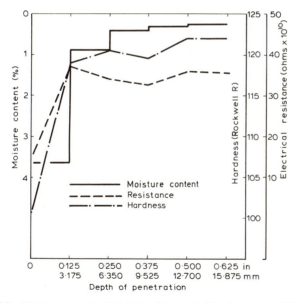

Figure 4.8 Moisture content, electrical resistance and hardness vs radial distance from surface (nylon 66 extruded, moisturised rod) (courtesy Polypenco Ltd)

The hardness value generally increases (although not necessarily linearly) with the modulus of the polyamide, and decreases with increase in temperature, moisture content and $CH_2/CONH$ ratio. The speed and convenience of the test commend it for use in cases where correlation can be made with other properties usually more difficult to evaluate. As shown in *Figure 4.8*, provided that a section at right angles to the extrusion axis is available, hardness can be used to determine the moisture gradient and extent of penetration in extruded nylon rod subjected to a moisturising treatment.

Since both the indentation tests mentioned are virtually non-destructive, the hardness method of test is favoured in production inspection of injection-moulded or extruded parts.

### 4.2.1.5  Impact properties

The impact properties of a polyamide article depends on a number of factors—for example, temperature of the material, moisture content, speed of impact, stress concentration effects, and anisotropy. For this reason results from standard impact tests often quoted in manufacturers' data sheets are unlikely to be a reliable guide to practical performance of the material. For design work the recommendations given in BS 4618 Section 1.2, *Impact behaviour*, if followed, are considered to be a good basis for predicting service behaviour, and the reader should consult this document for further guidance. The recommendations stress the need to obtain multi-point data covering a range of test specimens and test conditions. Particularly important is the need to determine notch sensitivity of the material by carrying out impact tests on un-notched specimens and on specimens notched to different radii and depth.

Standard impact tests such as the Izod and Charpy tests of BS 2782 Methods 306A, D, and E, the drop impact test of BS 2782 Method 306B, and the tensile impact test of ASTM D 1822 all, in general terms, allow comparison of impact behaviour between different nylon types, and the same types subjected to different conditions. Generally the anticipated impact behaviour of a material deduced from these tests is confirmed by its subsequent behaviour in service. The tests are, however, used more frequently in specifying material quality. As might be expected the impact strength of a polyamide increases with temperature and moisture content. In the absence of any second-order transition, increasing stiffness and decreasing impact strength are exhibited with reduction in temperature below

## Properties of polyamides

ambient. Moisture and plasticisers in the polyamide partially offset this effect but there is no sudden onset of brittleness. Nylons incorporating fibrous fillers have the advantage of being less notch-sensitive than the unfilled nylon. The former also maintain their impact strength better than the latter when temperature is reduced. *Figure 4.9*[18] shows for standard specimens (50 mm long × 6 mm broad × 3 mm thick) the effect of temperature and radius of notch on the Charpy impact strength of dry nylon 66 and 33-per-cent glass-fibre-filled nylon 66. *Figure 4.10*[18] shows the effect of notch radius on impact strength of dry unfilled and 33-per-cent glass-fibre-filled nylon 66.

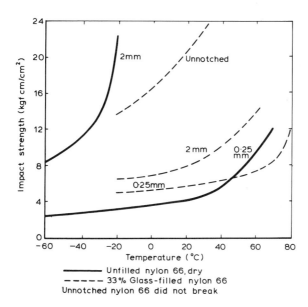

Figure 4.9  Impact strength vs temperature for nylon 66 showing effect of notch tip radius (courtesy ICI Ltd)

The notched impact strength of nylon 66 compares well with that of other commercially available engineering polymers tested under similar conditions. The values given in *Table 4.9* show the high ranking attained by nylon 66.

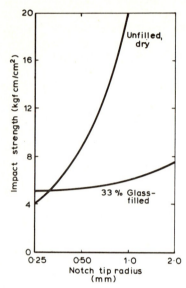

Figure 4.10 Impact strength vs notch tip radius for nylon 66 at 20°C (courtesy ICI Ltd)

The impact strength of polyamides increases with the average molecular weight, and this feature is observed even at low temperatures. For linear polyamides within the working range the impact strength increases with the $CH_2/CONH$ ratio. At the same time polyamides with large $CH_2/CONH$ ratio are less affected by exposure to low temperatures than those with small $CH_2/CONH$ ratio. Thus, for example, where designing for low temperature applications involving impact, a nylon 11 or nylon 12 may be satisfactory while the same design when temperatures above ambient are involved may require a nylon 66.

Polyamides are capable of withstanding repeated impact loading without fracture and they show up better in that type of test than other engineering plastics that appear to be more impact-resistant than polyamides in the Izod/Charpy type of test. *Table 4.10*[17] shows the difference between nylon 66 and cellulose acetate butyrate in this test. The data refer to an 18 mm (0.7 in) o.d. × 9 mm (0.35 in) i.d. roller struck on the outer surface by a free-falling 1.2 kg (2.7 lb) weight. The heights of fall required to cause a visible crack in one blow, or, for a repeated test, ten blows, are recorded. In this test moisture content of the nylon test specimen was 0.35 per cent.

### Properties of polyamides

**Table 4.9** TYPICAL NOTCHED IMPACT-STRENGTH VALUES FOR PLASTIC MOULDING MATERIALS

| Material | Notched impact-strength (ASTM D 256) | |
|---|---|---|
| | ft lbf/in | N m/m |
| ABS copolymer | | |
|   high-impact | 4–10 | 214–534 |
|   medium-impact | 2–4 | 107–214 |
| Polyacetal | | |
|   homopolymer | 1.4 | 75 |
|   copolymer | 1.2 | 64 |
| Nylon 66 | 1.5 | 80 |
| Polycarbonate | 2.5 | 133 |
| Polyethylene | | |
|   medium-density | 1–15 | 53–800 |
|   high-density | 1–15 | |
| Polypropylene | | |
|   homopolymer | 1.0 | 53 |
|   copolymer | 3.0 | 160 |
| Acrylic resin | | |
|   standard | 0.30 | 16 |
|   styrene copolymer | 0.35 | 19 |
|   high-impact | 1.20 | 64 |
| Polystyrene | | |
|   unmodified | 0.30 | 16 |
|   rubber-modified | 0.7–3.0 | 37–160 |
| Polyvinyl chloride | | |
|   unmodified | 0.7 | 37 |
|   rubber-modified | 15.0 | 800 |

**Table 4.10** REPEATED IMPACT TEST ON NYLON 66 AND A TYPICAL IMPACT-RESISTANT THERMOPLASTIC

| Material | Distance of fall | | | | Izod impact | |
|---|---|---|---|---|---|---|
| | One blow | | Repeated | | | |
| | in | mm | in | mm | ft lbf/in | N m/m |
| Nylon 66 | 35 | 890 | 30 | 760 | 2 | 107 |
| Cellulose acetate butyrate | 39 | 990 | 7 | 180 | 6 | 320 |

#### 4.2.1.6 Mechanical damping

Polyamides can, without failing, be subjected to higher dynamic loads than the majority of the other engineering plastics. This is due to the high capacity of the material to absorb energy. In rapid cyclic stressing this results in superior vibration damping. When the frequency of application of stress on a component exceeds a critical value (which depends on the polyamide molecular weight and other factors) heat is generated in the bulk of the material, and ultimately failure occurs due to excessive heat build-up. In the design of rotating components such as gears and propellers vibration damping is normally an asset, but under certain conditions allowance must be made to avoid the possibility of excessive heat build-up during working. Apart from modifying the section to reduce vibration, metallic inserts are often used to conduct away the heat generated in the component used. The damping capacity of a polyamide increases with temperature and moisture content. The degree of damping can be deduced by observing the shear modulus, and conveniently measured by evaluating the mechanical loss factor, tan $\delta$. Shear modulus decreases with increasing temperature, the reduction being marked when the region of high damping is reached. Similarly shear modulus decreases with increase in moisture content, which also has the effect of displacing the region of high damping towards lower temperatures. *Figure 4.11*[19] illustrates these features for nylon 6.

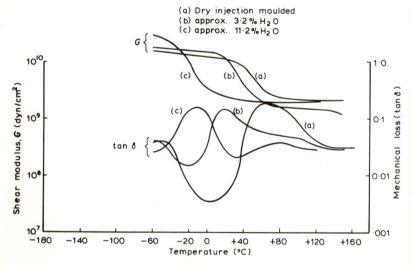

Figure 4.11  Shear modulus/mechanical loss factor vs temperature for nylon 6 at various moisture contents (courtesy Carl Hanser Verlag)

## 4.2.2 LONG-TERM PROPERTIES

Data on the properties of polyamides in the long term are essential for the proper design of components to achieve a specified performance throughout their lifespan. Two main conditions may be distinguished for stressed components. In the one the initial load applied to the component is maintained constant throughout its design life; in the other the component is loaded to a predetermined strain, which is held constant throughout the design life. In the former case there is continuous deformation of the component material with time, in the latter continuous decrease with time of the applied stress. These phenomena are known respectively as creep and stress relaxation. Components may be stressed statically or dynamically. In the latter case the stress rises from a zero or low value to a maximum and returns to the low value, the cyclic process being repeated usually at a constant frequency. As with metals this periodic type of stressing in polyamides can lead to dynamic fatigue, to be discussed later. Under conditions of static loading and using standard specimens the change of strain and stress over long periods can be determined, and the information presented in the form of creep and stress relaxation curves. The effect on these long-term properties of changes in the ambient temperature and humidity conditions is also important, and the curves and their derivatives are useful in component design for long-term performance. Data of this kind have been published in the design manuals of the main raw-material suppliers[16,18].

### 4.2.2.1 Creep

For the purpose of comparison of properties and to obtain values for design calculations, creep data may be presented graphically in various ways. The basic creep curves are obtained for standard test specimens by plotting total deformation against time for various loads. Tensile creep is the most easily measured, but creep may also be measured using other modes of deformation such as compression, tension or shear or a combination of these. Because of the extended time of observation semi-logarithmic paper is generally used to compress the time scale and accommodate a larger number of observations. Creep data for the different polyamide types are scattered throughout the accessible literature and direct comparison of creep performance is not always possible. *Figures 4.12*[18], *4.13*[20]

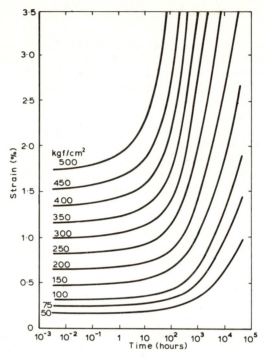

Figure 4.12 Tensile creep curves for dry nylon 66 at 20°C (courtesy ICI Ltd)

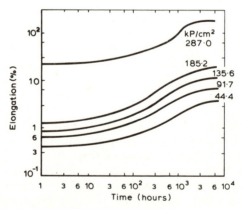

Figure 4.13 Tensile creep curves for nylon 6 conditioned 14 days at 65% RH, 20°C (courtesy Carl Hanser Verlag)

*Properties of polyamides*

and *4.14*[21] show respectively the uniaxial tensile creep curves for nylon 66, nylon 6 and nylon 12. *Figure 4.15*[22] compares in graphical form the flexural creep of nylon 66, nylon 6 and nylon 11.

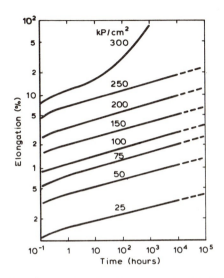

Figure 4.14   Tensile creep curves for nylon 12 conditioned at 50% RH, 23°C (courtesy *Kunststoffe*)

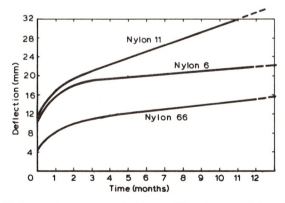

Figure 4.15   Flexural creep curves of nylons conditioned to equilibrium at 65% RH, 20°C (courtesy *Kunstoffe–Plastics*)

Using the basic creep data, the inter-relation between the parameters is often more clearly shown on the following derivation curves widely used by designers: (a) isometric stress vs log time curves, obtained by plotting for given strains the times required to reach the strains, against the stress applied; (b) isochronous stress vs strain curves, obtained by plotting for given times the stress applied for this time against the resultant strain at that time; (c) creep modulus (applied stress/creep strain) vs time curves for various strains. These curves enable the designer to determine the decay of stress with time at a given strain. The derivation of the curves of types (a), (b) and (c) from the basic creep curves is illustrated in *Figure 4.16*[18] and examples of curves of the types mentioned are shown in *Figures 4.17 to 4.19*.

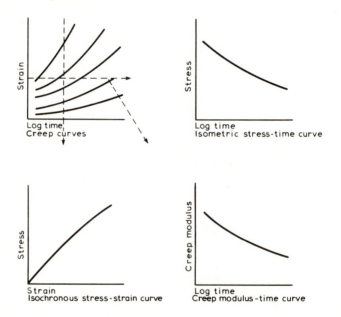

Figure 4.16  Derivation of isometric and isochronous stress curves from basic creep curves (courtesy ICI Ltd)

The creep behaviour of a given type of polyamide is affected by the temperature of the material, its moisture content, and the type and quantity of filler if any. The acquisition of data that may be used to forecast the creep behaviour of a polyamide is an arduous

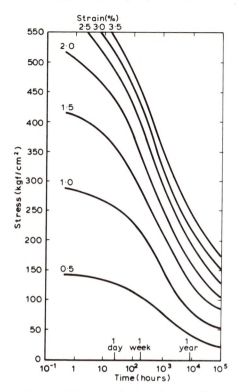

Figure 4.17  Isometric stress–time curves for dry nylon 66 at 20°C (courtesy ICI Ltd)

and time-consuming task. A selection of data has however been published by the main polyamide raw-material suppliers[16,18], covering the main commercial nylon types. The data include the effect on creep of temperature, humidity and fillers. There are also several guides and individual published papers referring to the creep data of polyamides[18,21,23,24].

Approximation and extrapolation methods are widely adopted by designers using existing data, in order to determine creep behaviour under specified conditions. The isochronous stress–strain curves of *Figures 4.20, 4.21*[18] and *4.22* illustrate the effect on creep of the variables mentioned. They show, respectively, the effect on creep of temperature for unfilled nylon 66, of moisture content for glass-filled

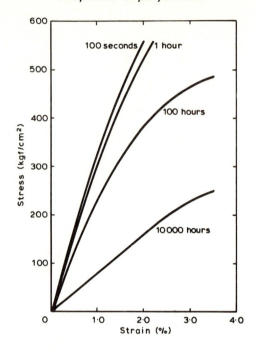

Figure 4.18 Isochronous stress–strain curves for dry nylon 66 at 20°C (courtesy ICI Ltd)

nylon 66 and of glass-fibre filler for nylon 6. *Table 4.11* summarises the effect of variables on the creep of polyamides.

Table 4.11 EFFECT OF VARIABLES ON CREEP IN POLYAMIDES

| Increase in variable | Effect on creep |
| --- | --- |
| Temperature | Increases |
| Moisture | Increases |
| Crystallinity | Decreases |
| $CH_2/CONH$ | Increases |
| Reinforcing fillers | Decreases |

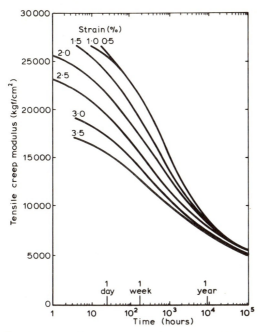

Figure 4.19  Tensile creep modulus–time curves for dry nylon 66 at 20°C (courtesy ICI Ltd)

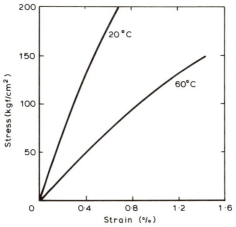

Figure 4.20  Effect of temperature on isochronous stress–strain curves for dry nylon 66, 100 seconds (courtesy ICI Ltd)

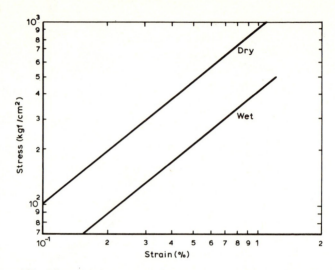

Figure 4.21  Effect of moisture on isochronous stress–strain curves for 33% glass-filled nylon 66 at 20°C, 100 seconds (courtesy ICI Ltd)

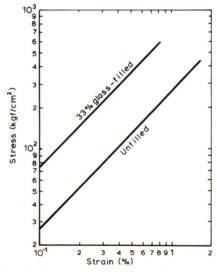

Figure 4.22  Effect of filler on isochronous stress–strain curves for dry nylon 66 at 20°C, 100 seconds (courtesy ICI Ltd)

## Properties of polyamides

### 4.2.2.2 Stress relaxation

Information on the decay of stress is required to predict long-term effects on polyamide engineering components such as locking screws originally stressed to a predetermined torque, closures intended to maintain tightness, and bushings assembled with an interference fit. While a set of stress-relaxation curves corresponding to the creep curves mentioned would be more useful for predicting the behaviour of such strained components, in practice results can be obtained from the corresponding family of isometric stress curves at various strains for the polyamide type of the component material, and for the conditions of service.

### 4.2.2.3 Recovery from creep

In many applications the load is applied intermittently to a component. In these cases, in the period between load applications, the strain tends to decrease, the degree of recovery being determined by the duration of the load and by the ratio of time on load to time off load.

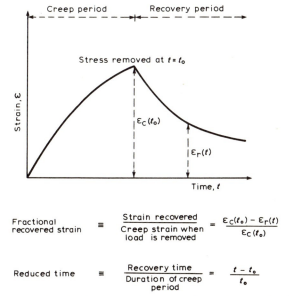

$$\text{Fractional recovered strain} \equiv \frac{\text{Strain recovered}}{\text{Creep strain when load is removed}} = \frac{\varepsilon_c(t_o) - \varepsilon_r(t)}{\varepsilon_c(t_o)}$$

$$\text{Reduced time} \equiv \frac{\text{Recovery time}}{\text{Duration of creep period}} = \frac{t - t_o}{t_o}$$

Figure 4.23  Graphical definition of creep coordinates (courtesy ICI Ltd)

Recovery data are conveniently presented as fractional-recovery/reduced-time curves, where

$$\text{Fractional recovery} = \frac{\text{Strain recovered}}{\text{Creep strain immediately before load removal}}$$

$$\text{Reduced time} = \frac{\text{Recovery time}}{\text{Duration of preceding creep period}}$$

The significance of these coordinates and their derivation from

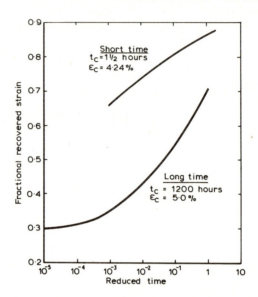

Figure 4.24  Recovery from creep in tension for dry nylon 66 (courtesy ICI Ltd)

strain–time curves is illustrated in *Figure 4.23*[18]. Data on recovery of nylon 66 from tension creep of two durations are shown graphically in *Figure 4.24*[18].

### 4.2.2.4  Long-term strength (static fatigue)

If the original stress level is sufficiently high a polyamide part subject to uniaxial tension will creep and ultimately fail due to necking rupture, if local yielding occurs, or brittle fracture without necking. Unfilled nylons commonly fail in a ductile manner; filled nylons can

## Properties of polyamides

fail through brittle fracture. The whole creep process culminating in failure is sometimes termed static fatigue. Static-fatigue curves are therefore extensions of the creep curves taken to the point of failure. *Figures 4.25, 4.26* and *4.27*[18] show long-term strength curves for nylon 66. These are isometric curves showing the effect of original stress on failure time for dry nylon 66, nylon 66 moisture-conditioned in various ways, and nylon 66 glass-fibre filled. The effect of environment on the fatigue life of polyamides is important.

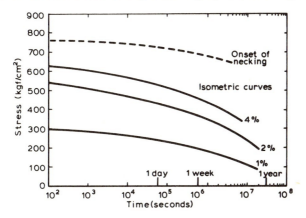

Figure 4.25  Long-term strength under constant load for dry nylon 66 at 20°C (courtesy ICI Ltd)

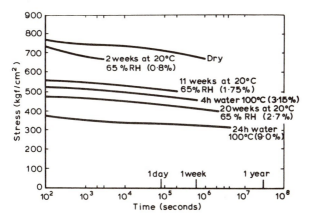

Figure 4.26  Long-term strength under constant load for nylon 66 at 20°C, moisture conditioned in various ways (courtesy ICI Ltd)

In environments other than dry or moist air (for example, organic liquids or aqueous salt solutions) continuously stressed parts may exhibit stress-crazing or stress-cracking, and hence fail at times much shorter than the fatigue life in the usual environments. A useful dissertation on static fatigue and experimental methods for studying fatigue behaviour is given in a paper by Gotham[25].

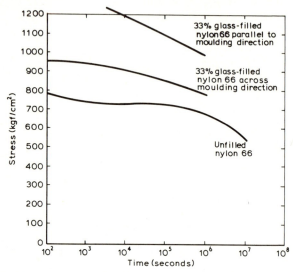

Figure 4.27 Long-term strength under constant load for glass-filled and unfilled nylon 66 at 20°C (courtesy ICI Ltd)

### 4.2.2.5 Dynamic fatigue

As with metallic materials, the application of a repeated cyclic stress to a thermoplastics material can ultimately lead to failure at a stress level lower than that determined from short-term static tests. This phenomenon is termed dynamic fatigue. It can occur when rotating or vibrating parts such as propellers or gears, fabricated from polyamides, are subjected to prolonged cyclic stresses. The number of cycles for failure of a stressed part depends not only on the stress level but also on the temperature of the part, its moisture content, the crystallinity of the material and the frequency of the applied stress. At frequencies greater than about 300 cycles/minute, the alternating strain energy tends to generate internal heat in the part, particularly at those temperatures where high damping occurs. This effect favours early failure due to thermal softening. Failure can also occur through crack propagation. Irreversible changes such as crazing and microvoiding may also be considered as failure, but for the purpose of determining failure time these terms would have to be quantitatively defined. Since damping occurs only in the amorphous regions of a semi-crystalline polymer such as a polyamide, fatigue due to thermal softening resulting from damping decreases with increase in the crystalline/amorphous ratio of the material. The effect on the dynamic fatigue of polyamides of the parameters mentioned above is discussed by Riddell et al[26] and Oberbach[27]. The latter also gives experimental data for nylon 66 and 6. Due to its high strength and excellent deformation properties, nylon is considered to be an excellent matrix material for glass-fibre reinforcement, and addition of this fibre increases the stress level required to cause failure by dynamic fatigue. The mechanism of fatigue failure in glass-reinforced plastics is discussed in a paper by Dally[28], who concludes that fatigue is largely initiated by a de-bonding process. Short fibres were found to be superior to long fibres in delaying fatigue.

Fatigue data on polyamides are usually obtained using standard test specimens. The stress mode may be tensile, compressive, shear, or combinations of these. *Figure 4.28*[16] shows fatigue curves for

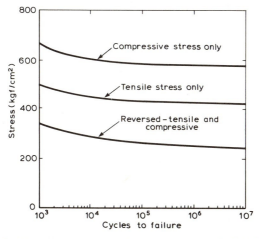

Figure 4.28 Tensile and compression fatigue for nylon 66 at 20°C, 2.5% moisture, 1200 cycles/minute (courtesy Du Pont de Nemours)

nylon 66 in three stress modes. Such fatigue data can only serve as an approximate guide to designers, and data obtained from simulated service testing under anticipated environmental conditions is much to be preferred.

### 4.2.3 ORIENTATION AND ITS EFFECTS

The common methods of processing polyamides, i.e. injection moulding and extrusion, result in some degree of molecular orientation in the material. In fact any process involving flow or shear will produce orientation of the crystalline and spherulitic structure, and this can occur either in the melt or in the plastic condition. While orientation occurs naturally or incidentally in most fabrication processes, it is sometimes possible to increase the orientation by using suitable process conditions with the object of increasing strength in the direction of orientation. In this way the strength of extruded nylon film can be controlled to a certain degree. With film also it is possible to orient biaxially in directions at right angles to each other, thus further increasing the overall strength of the film.

In injection moulding the direction of orientation in the vicinity of gates can be established, but in the interior of the moulding it is not usually possible to predict the direction of flow and hence the orientation, particularly with moulds of complex geometry. In this connection it is worth noting that many tensile test specimens used in determining standard properties of materials are end-gated, and the strength along the axis (i.e. direction of flow) is greater due to orientation than it would be for the corresponding test specimen with isotropic properties.

The situation in extrusion is somewhat different, since the melt viscosity has a greater influence on the orientation than is the case with injection moulding. The pattern of orientation also is symmetrical about a central axis. A high melt viscosity, resulting in high shear, favours orientation in the direction of flow while a low melt viscosity, which may result from an increase in melt temperature, reduces orientation. Considering these effects in relation to the extrudate section it can be seen that, as the material leaves the die, the highest degree of orientation should occur near the outer surface of the section, which cools faster, with increase in viscosity, than the interior. Orientation decreases towards the centre of the section. The net result is a variation of properties (discussed below) due to orienta-

## Properties of polyamides

tion across the section. The effect of flow orientation on short glass-fibre filled polyamides is even more marked than for the unfilled material since, in addition to molecular orientation, alignment of the short fibres along the flow direction occurs, resulting in considerably increased strength in this direction. The same considerations regarding the effect of melt viscosity on orientation hold with fibre-filled polyamides as with unfilled. With very high shear rates, stiff or brittle fibres can actually be fractured and the resulting length reduction tends to reduce the strength of the composite in the regions of high shear.

The principles governing molecular orientation in injection moulding and methods for determining orientation are discussed in a paper by Wiegand[29]. The effect of molecular and fibre orientation on material properties is briefly summarised below.

Tensile strength of an isotropic material is increased if it is subjected to orientation. The strength is increased in the direction of orientation and is proportional to the degree of orientation. In an orientated specimen impact strength is weaker in the direction of orientation, and stronger at right angles to this direction, than in the corresponding isotropic specimen.

Greater impact strength is also associated with a higher melt temperature, since dis-orientation tends to increase with rise in temperature.

Flexural strength is greater in the direction of orientation. Dis-orientation effects operate if higher melt temperatures are used during processing of the material, and increase in strength due to orientation tends to diminish. Uniaxial creep of a polyamide decreases with increase in orientation along the same axis. Decrease in melt temperature during processing increases orientation along the line of flow and decreases the creep susceptibility of the moulded or extruded material. The deflection temperature of polyamides that have been subjected to molecular orientation in processing is lower than that of the corresponding isotropic material. This is due to the tendency for the orientation stresses to relieve themselves as temperature rises, causing deflection movements not directly due to the normal deflection temperature response of the isotropic material.

With glass-fibre filled polyamides it is found that, in addition to the effects described above for unfilled materials, there is a lower coefficient of thermal expansion and a lower mould shrinkage in the direction of fibre orientation. In some cases residual stresses in polyamides, due to orientation, can be released when the material is exposed to chemical environments such as certain aqueous salt solutions. This often results in cracking or crazing of the stressed parts.

### 4.2.4 FRICTION, WEAR AND BEARING PROPERTIES

Where components are designed to move relative to, and in contact with, surfaces of other components, low friction and good wear and bearing properties are generally required. Fortunately most polyamides have a useful combination of these properties and their significance will be discussed in the sections following. These properties are of great interest to the designer and engineer, and the discussion will be at a fairly practical level with the theoretical aspect of the subject kept to a minimum.

#### 4.2.4.1 Friction

For dry sliding conditions, standard quoted values of coefficient of friction of plastics materials rank polyamides higher than most other thermoplastics, as shown in *Table 4.12*[30]; engineering experience confirms this ranking. Published tables of static or dynamic coefficients do little more than permit a first selection of materials for intended end-use in sliding friction applications. More informative values of friction coefficient valid for the conditions of use are commonly obtained from test rigs that simulate the service use of the component. This practice is recommended since the coefficient of friction is considerably influenced by many variables such as load, relative speed at interface, surface finish, composition of the test face, temperature and humidity, all of which may be controlled in the test.

Friction plays a dominant role in wear at the interface but many other factors contribute. The resistance of frictional forces to movement primarily generates heat and, since unfilled plastics materials have low thermal conductivities, the limitation to their use in sliding applications may be their inability to perform at high temperature because of unacceptable heat deformation. This property is expressed in the limiting *PV* rating, to be discussed later.

From the engineering viewpoint friction and wear are normally inseparable except in single-pass operation, i.e. where the wear path is traversed only once. Two conditions can be distinguished, namely low load with high speed (as obtained for instance in bearings or plastic bushes) and high load with low speed (such as is encountered

## Properties of polyamides

with static bearing pads designed to accommodate small movement of heavy structures). More is known of the performance of polyamides operating under the former condition, particularly as plastic bearings.

Table 4.12 COEFFICIENT OF FRICTION OF VARIOUS PLASTICS MATERIALS (courtesy Royal Society)

| Materials | Plastic on plastic | Plastic on steel | Steel on plastic |
|---|---|---|---|
| Polythene | 0.25 | 0.25 | 0.3 |
| Polyvinylchloride | 0.4–0.45 | 0.35–0.40 | 0.4–0.45 |
| Polymethylmethacrylate | 0.4–0.6 | 0.5 | 0.45–0.5 |
| Polystyrene | 0.4–0.5 | 0.4–0.5 | 0.4–0.5 |
| Nylon 66 | 0.3 | 0.25 | 0.3 |
| Polytetrafluoroethylene | 0.04 | 0.04 | 0.1 |

Modern theories of friction still recognise the basic laws first enunciated by Amonton in 1699. The first law, which states that the frictional force of one body sliding over another is proportional to the load normal to the interface, can be expressed as

$$\mu = F/W$$

where

$\mu$ = coefficient of friction
$F$ = frictional force
$W$ = normal load

The second law states that the friction is independent of the area of contact of the two bodies. It is now known that the real area of contact of two bodies, one supporting the other, is very much less than the apparent area of contact and that the load is supported on a small number of asperities at the interface. On first contact of the two bodies these asperities deform by plastic flow until the applied load is supported in an equilibrium position such that the real area of contact can be expressed by the equation

$$A = W/P_m \qquad (4.9)$$

where

$A$ = real area of contact
$W$ = load
$P_m$ = yield pressure of material

It is now believed[31] that in such cases the mating surfaces, plastically deformed by the loading applied, adhere at the asperities. If movement parallel to the interface is desired the force to be overcome is the friction, and this is equal to the shear force necessary to break the points of contact. Thus

$$F = AS \qquad (4.10)$$

where

$S$ = shear strength of the material

Equations (4.9) and (4.10) may be combined as

$$F = WS/P_m \qquad (4.11)$$

where $\mu$ may be identified with the term $S/P_m$. Equation (4.11) strictly applies to metals, which generally exhibit true plastic deformation above their yield point. For polymers (which generally possess a degree of visco-elasticity) the area of true contact is given by the relationship

$$A \propto W^n$$

where $n < 1$. The coefficient of friction of polymers would therefore tend to decrease with increase in load, and this has been experimentally observed.

Friction changes only slightly with temperature for polyamide surfaces operating under dry sliding conditions, provided the maximum temperature is not too near the melting point. *Table 4.13*[31] gives values of the coefficient of friction of nylon 66 at different temperatures obtained using a modified Bowden-Leben machine. For dry sliding friction the finish of the contacting surfaces influences the friction developed and, as expected, a good surface finish shows a lower friction than one with a poor finish.

Increase in moisture content increases the friction of polyamides and, since the layers near the surface are first and most affected, the frictional changes effected by moisture are usually quick to respond to environmental conditions.

Investigations into plastic-metal friction have shown that with continuous sliding contact plastic material is transferred to the metal surface—and vice versa, to a certain extent. If a stable adherent film is formed, a constant-friction regime is set up and a constant wear rate is observed in the plastic until degradation of the film sets in. This is analogous to the boundary lubricant films encountered in conventional oil lubrication. With polyamides working in dry sliding conditions, adherent nylon films have been observed and studied particularly with regard to their permanence.

While adherent films may be derived from the parent polyamide,

## Properties of polyamides

**Table 4.13** EFFECT OF TEMPERATURE ON THE COEFFICIENT OF FRICTION OF NYLON 66 (courtesy Clarendon Press)

| Temperature (°C) | Coefficient of friction | | |
|---|---|---|---|
| | Machined nylon on machined nylon | Machined nylon on steel | Steel on machined nylon |
| 20 | 0.46 | 0.43 | 0.33 |
| 60 | 0.45 | 0.43 | 0.36 |
| 100 | 0.46 | 0.54 | 0.37 |
| 120 | 0.46 | 0.54 | 0.45 |
| 140 | | 0.53 | |
| 160 | | 0.57 | |
| 180 | | 0.60 | |

much research has gone into developing formulations incorporating additives that preferentially plate out on the metallic surface to form permanent adherent films resistant to degradation under high-load and high-speed conditions. These formulations are used for the high-$PV$ materials now being produced for plastic bearings.

The additives may be inorganic materials of plate-like structure such as graphite or molybdenum disulphide with proven performance as lubrication aids. The transfer film may be the additive itself, or the additive together with wear debris from the polymer matrix. More recent practice has been to incorporate polymeric materials of low intrinsic frictional coefficient and deformational

**Table 4.14** KINETIC COEFFICIENT OF FRICTION AND WEAR RATE OF POLYAMIDES SLIDING ON STEEL (courtesy Polypenco Ltd)

Test period = 24 hours
Surface finish on steel = $2\,\mu m$
Surface pressure = $0.5\,kP/cm^2$
Surface velocity = $0.6\,m/s$
Approximate surface temperature = 40°C

| Polyamide | Coefficient of friction | Wear rate ($\mu m/h$) |
|---|---|---|
| Nylon 66 | 0.35–0.42 | 0.09 |
| Nylon 6 | 0.38–0.45 | 0.23 |
| Cast nylon 6 | 0.36–0.43 | 0.10 |
| Nylon 6.10 | 0.36–0.44 | 0.32 |
| Nylon 11 | 0.32–0.38 | 0.80 |
| Nylon 66 + 2.5% $MoS_2$ | 0.30 | 0.05 |
| Cast nylon 6 + 1.5% $MoS_2$ | 0.30 | 0.06 |
| Nylon 66 + 8% HDPE | 0.19 | 0.10 |

properties superior to the polyamide matrix. Polyethylene and polytetrafluoroethylene have been used in this way. *Table 4.14*[32] shows the kinetic coefficient of friction and wear rate of various filled and unfilled polyamides against manganese-chrome steel.

### 4.2.4.2 Wear of polyamides

Of the several wear mechanisms to which polyamides are subject in service, the two most important are abrasive wear and adhesive or sliding wear. In continuous-sliding applications, however, a third type classed as fatigue wear tends to take over from abrasive wear as the contacting surfaces become smoother. This type of wear is characterised by local detachment of material as a result of cyclic stress variations.

Another type, cavitation and erosive wear, can occur in components subjected to high-speed movement in fluids. Because of higher wear resistance, polyamides have been successful in replacing metallic materials such as bronze in these applications. It has been shown, for example, that propeller blades made in monomer-cast nylon used in small coastal and naval vessels suffer far less from cavitation and erosion than the bronze blades they replaced. The resistance of the nylon to sea-water corrosion is in this application an additional factor in its favour. Polyamides have also been used successfully in the lining of chutes and in belt conveyors and conveyor buckets subject to erosion from hard impacting particles such as coal or mineral ores. Where traditional linings from materials such as steel have been replaced by nylon, the component life has been considerably prolonged. Reports from operators of bucket conveyors handling abrasive materials confirm that replacement of steel buckets by nylon monomer-cast buckets has resulted in a three-fold increase in bucket life.

The subject of abrasive wear of polymers has been treated in depth by Lancaster[33]. This author describes experimental work on two-body abrasion of polymers on rough metal surfaces, abrasive papers and metal gauze, and attempts to give a general picture of the wear processes involved. In the case of both two-body abrasion (which has been defined as wear by displacement of material from surfaces in relative motion caused by the presence of hard protuberances) and three-body abrasion (which is wear caused by the presence of hard particles either between the surfaces or embedded in one of them), polyamide surfaces show excellent performance. *Figure 4.29*,

## Properties of polyamides

taken from Lancaster's paper[33], shows comparative wear rates of polymers and metals on abrasive paper and smooth steel. Note that these data refer to single traversals of lubricated surfaces. Nylon shows the highest wear ranking (i.e. lowest rate) of all the polymers examined; this has been attributed to its combination of high elasticity, plasticity, and good fatigue properties compared with the other more rigid polymers. In cutting-type wear processes, such as are encountered with abrasive paper, it would seem from *Figure 4.29*

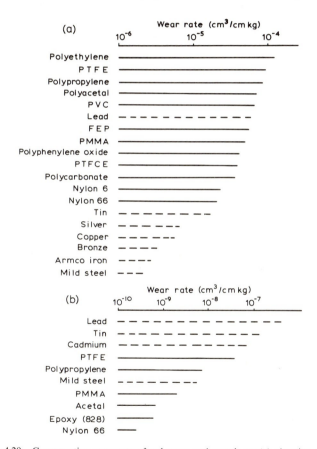

Figure 4.29  Comparative wear rate of polymers and metals on (a) abrasive paper (Grade 100D) and (b) smooth steel (0.15 µm CLA); single traversals lubricated (from *Wear*; Crown copyright, reproduced by permission of the Controller, Her Majesty's Stationery Office)

that the wear resistance of metals is superior to that of polymers but nylon still shows the best performance of the polymers examined.

Field experience of nylon components in engineering applications has shown that the wear resistance in three-body abrasion is high. In many industries such as mining and cement manufacturing, where dust-laden atmospheres exist alongside moving machinery, wear-resistance of components in three-body abrasion conditions is called for. It cannot yet be said that standard abrasion tests on materials are the best basis of selection for particular applications, and until the wear processes are more thoroughly understood there is no good substitute for practical service trials.

As with metallic materials, the adhesive or sliding type of wear has been studied extensively for polymers, with the object of establishing a reliable theory for the wear process in order to predict service behaviour and the useful life of components working under such conditions of wear. In fluid lubrication under either hydrodynamic or boundary conditions, the theory established for metallic surfaces generally holds good for plastic surfaces. Unlike metals, polymers and particularly polyamides are capable of working effectively in dry conditions or after only an initial smear of lubricant. On this basis polyamides are often chosen in sliding mechanisms or bearing applications where it is impractical or inconvenient to lubricate the surfaces from an external source.

Very many types of wear-test machines using different techniques have been used to assess the amount of wear in polymers under sliding friction. Commercially available machines are the SAE Twin Disc (Amsler type) and the Denison T62. A useful classification of these machines and a summary of the literature concerning wear testing has been produced by the British Hydromechanics Research Association[34]. A common type of wear-tester much favoured by American workers uses the plastic under test as a bushing, housed in a cylinder, that can be loaded against a rotating metal shaft. The volume of wear particles removed depends on the load supported and the distance one surface travels relative to the other. In this type of tester, under constant temperature conditions, a wear factor characteristic of the plastic can be evaluated thus:

$$v = KFVT$$

where

$v$ = volume of wear particles removed (in$^3$ or m$^3$)
$V$ = sliding velocity (ft/min or m/s)
$F$ = applied load (lbf or N)

$T$ = time (h)

$K$ = wear factor $\left(\dfrac{\text{in}^3/\text{min}}{\text{ft lbf/h}} \quad \text{or} \quad \dfrac{\text{m}^3/\text{s}}{\text{N m/h}}\right)$

$K$ may be stated in Imperial or SI units as shown. Depending on the system used, the appropriate units for the factors in the expressions must be used in evaluating $K$. The load applied normal to the axis may also be converted to a pressure, using the projected area for the bushing, and the derived expression for wear becomes

$$R = KPVT \qquad (4.12)$$

where $P = F/A$ (lbf/in$^2$ or N/m$^2$) and $A$ = projected area (in$^2$ or m$^2$); $R$ is then the radial wear (in or m).

The advantage of using the above relationship is that the wear rate can be directly related to the pressure velocity conditions ($PV$) and therefore also to the bearing performance of the material (see section 4.2.4.4). Values for various bearing materials under one or both of two $PV$ conditions taken from the data of Pratt[35] are shown in *Table 4.15*. The performance of nylon compositions is shown to compare well with that of compositions based on other plastic matrices. The improvement in wear performance produced by the

**Table 4.15** SPECIFIC WEAR RATE OF COMMERCIAL DRY BEARING COMPOSITIONS

| Composition | Radial wear rate | | | |
| --- | --- | --- | --- | --- |
|  | at *3000 $PV$ | | at *6000 $PV$ | |
|  | μin/h | μm/h | μin/h | μm/h |
| Nylon 66 | 13–48 | 330–1220 | — | — |
| MoS$_2$-filled nylon 66 | — | — | 8.9 | 226 |
| MoS$_2$-filled sintered nylon | — | — | 2.4 | 61 |
| Oil-filled sintered nylon | — | — | 1.2 | 30 |
| Polyacetal | 9.1 | 231 | 78 | 1981 |
| Glass-fibre filled polycarbonate | — | — | 86 | 2184 |
| MoS$_2$-filled epoxy | — | — | 91 | 2311 |
| 15% glass-fibre filled PTFE | — | — | 26 | 660 |
| Bronze-filled PTFE | — | — | 63 | 1600 |

* $PV$ values are given here in lb in$^{-2}$ ft$^{-1}$ min$^{-1}$ (1 lb in$^{-2}$ ft$^{-1}$ min$^{-1}$ = 0.0286 N m$^{-1}$ s$^{-1}$)

incorporation of molybdenum disulphide in the nylon composition is also shown.

A useful method of wear testing, termed the cross-cylinder method, has been found by the author to be convenient and rapid for ranking plastic materials in terms of their wear rates. In this method cylindrical sections of the test material are loaded against a rotating shaft whose axis is at right angles to the axis of the test piece. Apart from test-material type, the load, sliding velocity, shaft material, shaft finish and ambient conditions can be controlled. The elliptical wear scar that develops on the test piece is proportional to the volume wear, which can be calculated from measurements of the major and minor axes of the scar. For a fixed load and sliding speed the volume wear is shown for most plastics to be linear with time, after a relatively short running-in period, and the constant wear-rate obtained characterises the wear behaviour of the material. It has been found that if the load and speed are chosen correctly linear plots from which wear rates can be calculated can, for the majority of plastics, be obtained in a few days.

*Figure 4.30* shows typical plots for nylons, obtained in the author's laboratory. *Table 4.16* lists wear rates, obtained by the author using the cross-cylinder method, for various plastics material of potential application in bearings. The effect of filler on the ranking of the

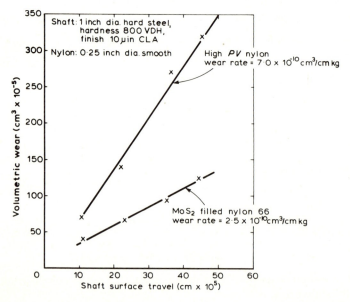

Figure 4.30 Volume wear rate for nylons; cross-cylinder method (courtesy Polypenco Ltd)

nylons in this test is noteworthy. The results may be used to make an initial selection of material for a specific wear application. Correlation of cross-cylinder wear data with service performance of any thermoplastic is not guaranteed, and it is preferable to confirm the selection by comparison with wear performance on rig tests where the service conditions are simulated.

Table 4.16 SPECIFIC WEAR RATE OF PLASTICS BEARING MATERIALS: CROSS-CYLINDER METHOD

| Material | Specific wear rate $(cm^3 cm^{-1} kg^{-1} \times 10^{-10})$ |
|---|---|
| Extruded nylon 66 | 4.0–4.8 |
| $MoS_2$-filled nylon 66 | 1.6–3.0 |
| Monomer cast nylon 6 | 1.6–2.4 |
| Monomer cast nylon 6/12 copolymer 90:10 | 1.6–2.1 |
| Monomer cast nylon 6/12 copolymer 80:20 | 2.8–6.2 |
| Polyacetal homopolymer | 0.1–0.2 |
| PTFE-filled polyacetal homopolymer | 0.6–0.7 |
| Unfilled PTFE | 55–93 |
| Mica-filled PTFE | 0.7–1.0 |
| High molecular weight polyethylene | 0.2–0.3 |

In assessing the wear performance of plastics bearings, a distinction must be made between wear rate and wear lifetime. As already stated, after initial run-in wear rate can be determined in the short term using one or other of the standard wear testers mentioned. In practice, if wear tests are continued for long periods it has been shown that for some plastics, including polyamides, the constant-wear-rate regime can be terminated by the onset of rapid wear arising from the breakdown of transfer films and other factors. This defines the wear lifetime of the bearing.

To characterise completely the wear performance of a bearing material, an evaluation of wear lifetimes for different $PV$ conditions of the bearing would be required. In practice, as with creep testing, a few $PV$ conditions are chosen for which the wear lifetime is evaluated. Wear lifetimes for other $PV$ conditions may be obtained by interpolation.

Evaluation of wear lifetimes using the method of loaded bearings usually takes an inconveniently long time, but accelerated methods of determination are not applicable in this method. The cross-cylinder method of wear-rate ranking is useful in evaluating the potential of polyamides incorporating fillers designed to reduce wear; it has not been used to determine wear lifetime, and is probably unsuitable for long running-periods because of inaccuracies in measuring wear scars of deep profiles.

### 4.2.4.3  Bearing properties

Polyamides are now widely used for the load-bearing elements in bearings such as plain journal bearings, thrust bearings and the like, and also for cages for ball or roller bearings. In load-bearing elements the capacity of polyamides to support high loads at substantial sliding velocities with low wear rates of the polyamide itself, combined with freedom from corrosion and noise during running, has been the main factor that has led many designers to replace traditional bronze or white-metal bearings with types incorporating polyamides. The art has advanced to the extent that at least one national authority now has a standard covering thermoplastic bearing materials. The German VDI (Verein Deutscher Ingenieure) has issued a draft document (VDI 2541) giving general information and design recommendations on unfilled thermoplastic material to be used in bearings. The intention is to prepare a further document covering filled thermoplastics for bearings. The British Standards Institution has a committee considering plastics bearing material and design standards.

### 4.2.4.4  Bearing materials evaluation

The load-bearing capacity of a thermoplastic in the form of a bearing is generally expressed as a limiting $PV$ value, that is, the maximum value of the product of the pressure on the bearing and the shaft velocity relative to the bearing, which permits equilibration of temperature and friction. The limiting $PV$ value is not specific for the material, but varies with the size and geometry of the bearing, surface finish of the shaft and several other factors. If the bearing is run under $PV$ conditions in excess of the limiting value it will melt, deform, or show unstable frictional behaviour. Limiting $PV$ values are often determined using an accelerated test known as the step-test in which the load on the running bearing is increased stepwise in equal increments, at equal short-time intervals of about 45 minutes. The friction torque and temperature are monitored throughout the test. The limiting $PV$ is taken as corresponding to the conditions

## Properties of polyamides

where the torque or temperature fail to come to equilibrium. O'Rourke[36] has given details of this test. For normal working limits, $P$ can be plotted against $V$ to show the locus of all values of

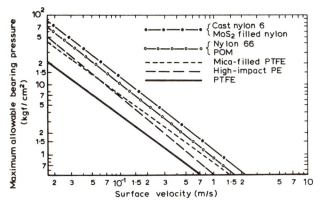

Figure 4.31 Typical $PV$ plots for plastics bearing materials (unlubricated continuously operating at 50% RH, 23°C)

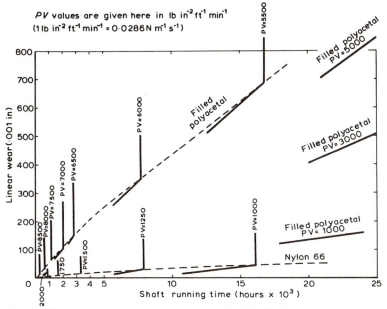

Figure 4.32 Bearing performance of nylon 66 and PTFE-filled polyacetal at various $PV$ conditions (courtesy Polymer Corporation)

these variables giving a product equal to the limiting value. If log-log coordinates are used the plot is nearly linear. *Figure 4.31*[37] gives the *PV* plots of some commercial thermoplastics materials and shows the high rating enjoyed by the nylons.

Most workers in this field agree that the limiting *PV* value is merely a measure of the bearing's ability to resist increase of temperature up to the deformation region of the material under the operating conditions and in the particular configuration of the bearing. By itself it can never predict the long-term performance of the material since no measurement of wear is made. To determine long-term performance, wear tests at several *PV* conditions below the limiting *PV* values must be carried out and the wear rate and the wear lifetime (i.e. time to rapid wear) determined for each *PV* value. *Figure 4.32*[38] shows wear rate and wear lifetime plots at various *PV* conditions for a commercial polyamide and a filled polyacetal.

#### 4.2.4.5 Effect of fillers

The bearing properties of polyamides can be improved by incorporating certain fillers, often termed high-*PV* additives. The filler has two main functions, to reduce friction and wear rate and/or to increase the thermal conductivity of the material so that the temperature at the surface of the bearing is reduced. A direct consequence of lower friction in a bearing is a lower temperature at the interface.

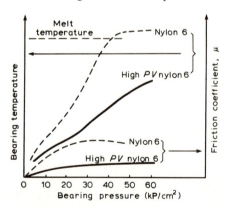

Figure 4.33 Effect of high *PV* additive on bearing temperature and friction coefficient of nylon 6 (courtesy Heinz Faigle)

*Properties of polyamides* 117

Alternatively, to obtain the same interfacial temperature a greater loading on the bearing can be used, i.e. a higher *PV* rating. *Figure 4.33*[39] shows the effect on coefficient of friction and bearing temperature of adding a high-*PV* additive to nylon 6.

The advantages of lower friction and surface temperature derived by the high-*PV* additive should not be offset by excessive wear rate or lowered mechanical properties of the product. The accepted formulations for high-*PV* material do not show these disadvantages to any marked degree.

## 4.3 Moisture absorption and its effect on mechanical properties

### 4.3.1 MOISTURE ABSORPTION

Polyamides as a class are more hygroscopic than most thermoplastics. Liquid water or water vapour can be absorbed from the surroundings in a proportion approaching ten per cent by weight of polymer, depending on the type. This property raises problems in processing and design of components made from polyamides, for not only are most properties considerably affected by the water absorption but dimensional changes may occur that can affect the functioning of components.

For processes such as injection moulding or extrusion, it is usually necessary to use the resins dried to a low specified moisture content. The raw material must be supplied in sealed containers, which should not be opened until just before processing operations. For designing in polyamides, it is essential to know the effect of the absorption of water by this class of compound, but knowledge of the principles governing the absorption processes is required for a fuller understanding of the behaviour of components in service. Absorption can be considered from a kinetic or thermodynamic viewpoint; that is, by studying either the rate of the process, or the final equilibrium of the polyamide with the environment. These aspects are discussed in some detail in the following sections.

#### 4.3.1.1 Absorption kinetics

The rate of water absorption and desorption in polyamides is diffusion-controlled and strongly temperature-dependent. Theoretically-derived equations may be used with simplifying assumptions to evaluate the diffusion coefficients of water vapour passing through various types of polyamide under various conditions. For example, the following relationship has been found valid for polyamide in plate form:

$$c_t = c_s \frac{2.256}{s} \sqrt{(Dt)} \qquad (4.13)$$

where
$c_t$ = concentration of moisture in part after $t$ seconds (g/cm³)
$c_s$ = saturation concentration of moisture (g/cm³)
$s$ = plate thickness (cm)
$D$ = diffusion coefficient (cm²/s)

Equation (4.13) allows calculation of the moisture absorption of plates of polyamide if the saturation concentration and the diffusion coefficient are known for the polyamide type being investigated. The diffusion coefficient may be evaluated from equation (4.14):

$$q/Q = 1.128\omega Dt \qquad (4.14)$$

where
$q$ = moisture absorbed in $t$ seconds (g)
$Q$ = total moisture absorbed at saturation (g)
$\omega$ = ratio of surface (cm²) to volume (cm³)

In practice $D$ is obtained by plotting $q/Q$ against $\omega\sqrt{t}$. Linear plots are obtained for plates of different thicknesses, the slope being proportional to $\sqrt{D}$ whence $D$ can be evaluated. *Table 4.17* shows the diffusion coefficient and saturation-moisture content of three commercial polyamides.

**Table 4.17** SATURATION MOISTURE CONTENT AND DIFFUSION CO-EFFICIENT AT VARIOUS TEMPERATURES, OF SOME COMMERCIAL POLYAMIDES

|  | Polyamide type | | |
| --- | --- | --- | --- |
|  | 6 | 66 | *6.10* |
| Saturation moisture content (%), average values | 10 | 9.0 | 3.5 |
| Diffusion coefficient, $D$ ( $\times 10^8$ cm²/s) | | | |
| at 20°C | 0.5 | 0.12 | 0.05 |
| at 40°C | 1.6 | 0.50 | 0.24 |
| at 60°C | 5.0 | 2.2 | 1.20 |
| at 80°C | 17 | 10 | 6 |
| at 100°C | 55 | 40 | 30 |

## Properties of polyamides

Equation (4.13) can be utilised to determine the times required to condition nylon, in plate form of different thicknesses, to a specified moisture content at various temperatures. The relationships shown graphically in *Figure 4.34*[40] have considerable practical use in determining the conditions required for stabilising nylon components.

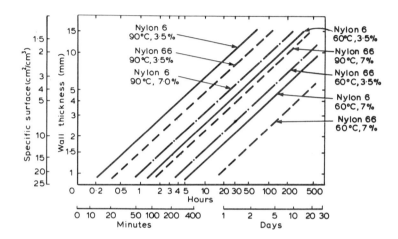

Figure 4.34  Time to condition nylon 6 and nylon 66 plate to 3.5% and 7% moisture content at 60°C and 90°C, as a function of section thickness; complete water immersion (courtesy Carl Hanser Verlag)

Like absorption, the drying out or desorption of water vapour from a moisture-containing polyamide is diffusion-controlled, and an equation similar to (4.14) can be derived for the process:

$$\frac{q}{V(C_E - C_A)} = \psi[\omega\sqrt{(Dt)}] \qquad (4.15)$$

where

$C_E$ = concentration of water in nylon at time $t$
$C_A$ = concentration at start of drying
$V$ = total volume
$\psi$ = shape factor
$\omega$ is as in equation (4.14)

### 4.3.1.2 Absorption equilibrium

The important factors in the establishment of equilibrium between a polyamide and a water-containing environment are:

(1) the relative humidity (RH) or partial vapour pressure of water vapour in the surroundings;
(2) the $CH_2/CONH$ ratio in the polyamide;
(3) the degree of crystallinity in the polyamide.

The value of equilibrium moisture absorption of a polyamide in moist air depends on the RH during exposure. It rises with increasing RH until the equilibrium moisture absorption at saturation is reached, at 100 per cent RH. This value represents the maximum uptake from moisture-saturated air, and will not be significantly increased by complete immersion of the polyamide sample in liquid water.

In practice, since the absorption process is diffusion-controlled and relatively slow, equilibrium conditions in nylon components are seldom attained. For components requiring dimensional stability, the designer should aim to take into account the service conditions to which the material may be exposed and ensure that the component is at least partially moisture-conditioned by pretreatment, to attain near-equilibrium with the surroundings so that further dimensional changes are minimised. Partial conditioning may be sufficient for parts always to be exposed to fluctuating humidities within the approximate range 20–70 per cent RH, since the comparatively rapid climatic changes that occur hardly affect the much slower diffusion process in the polymer, and dimensional change in the part is much reduced.

Complete moisture conditioning is usually necessary for parts exposed continuously to liquid water. Methods for conditioning polyamide components or shapes are given in Chapter 5.

The effect of the $CH_2/CONH$ ratio of the polyamide on the equilibrium moisture content is shown in *Table 4.18*[41]. From the values shown it is apparent that the greater this ratio the smaller the equilibrium moisture absorption. The degree of crystallinity of the polyamide has also a significant effect on the equilibrium moisture absorption and, as shown in *Table 4.19*[42] where the values for various forms of nylon 6 are compared, maximum moisture absorption decreases with increasing crystallinity.

## Properties of polyamides

**Table 4.18** EQUILIBRIUM MOISTURE ABSORPTION OF COMMERCIAL POLYAMIDES IN STANDARD ATMOSPHERE AND IN WATER (courtesy Carl Hanser Verlag)

| Polyamide | $CH_2/CONH$ ratio | Equilibrium moisture absorption (%) | |
|---|---|---|---|
| | | Standard atmosphere (65% RH, 20°C) | Water (20°C) |
| 6 | 5 | 3.0–4.2 | 9–11 |
| 66 | 5 | 3.4–3.8 | 7.5–9 |
| 6.10 | 7 | 1.8–2.0 | 3–4 |
| 11 | 10 | 1.1 | 1.6–1.8 |
| 12 | 11 | 1.0 | 1.5 |

**Table 4.19** MAXIMUM MOISTURE ABSORPTION OF NYLON 6 OF DIFFERENT DEGREES OF CRYSTALLINITY (courtesy Carl Hanser Verlag)

| Form of nylon | Percentage crystallinity | Equilibrium moisture absorption (%) |
|---|---|---|
| Stretched film | 15 | 12 |
| Standard plate | 27 | 11 |
| Plate of refined crystallinity | 35 | 9.5 |
| Extruded rod (100 mm dia.) | 43 | 8.9 |
| Cast rod, anionically polymerised (100 mm dia.) | 53 | 6.0 |

It has been established that the equilibrium moisture absorption of polyamides depends very little on the temperature of the medium. This is fortunate in moisture-conditioning where the process can be speeded up by treating the material in hot or boiling water, and this is indeed a common practice.

### 4.3.1.3 Graphical presentation of kinetic and equilibrium properties

Since the moisture absorption of polyamides is dependent on so many factors the description of the behaviour of the individual types is best described in graphical form. This practice is widely used by raw material suppliers and usually takes the form shown in *Figure 4.35*[43], where the absorption of the polyamide type for various

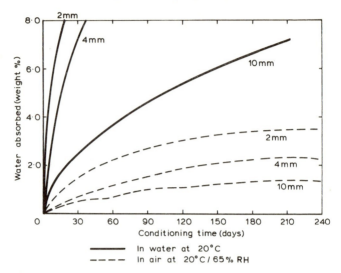

Figure 4.35 Moisture absorption vs conditioning time for varying thickness of injection-moulded nylon 6 plate (courtesy Carl Hanser Verlag)

thickness sections and at a given relative humidity is plotted against time.

### 4.3.1.4 Dimensional changes due to moisture absorption

The dimensional change of polyamide parts arising from absorption of moisture is of vital concern to their design and service. It has been shown that theoretically for nylon 6 and 66 a weight increase of 1 per cent due to moisture absorption amounts, if the distribution is uniform, to a maximum volume increase of 0.93 per cent. This corresponds to a linear dimension increase of about 0.3 per cent. These values are seldom realised in fabricated parts, since the distribution of moisture is not uniform throughout the section. More often the interior of the section is more or less dry, and a moisture concentration gradient exists in the outer layers.

The effect of this non-uniformity on the dimensions of nylon engineering components usually requires individual consideration. For example bearing bushes, if thin-walled, will increase on both the outside and inside diameters with moisture absorption, as the distribution tends to be uniform throughout the section. Assuming

## Properties of polyamides

free access of moisture to the outer and inner surfaces, thick-walled bushes or cylindrical sections will in the first instance increase on the outside diameter and decrease on the inside. When the water is uniformly distributed throughout the section the inside diameter will be shown to have increased. Restricted access of moisture to the outer surface, such as might occur in a sealed housed journal bearing, will result in appreciable decrease in internal diameter and close-down on the shaft since only the internal surface absorbs moisture.

*Figure 4.36*[44] shows a typical relationship between moisture absorption and linear increase. Here the linear expansion is shown to be only 0.15 per cent for 1 per cent moisture absorption. *Figure 4.37*[45] shows moisture absorption and corresponding linear increase related to conditioning time in air for nylon 6 mouldings.

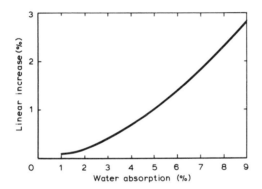

Figure 4.36  Linear increase vs water absorption for injection-moulded nylon 6 specimens, 120 mm × 10 mm × 4 mm at 20°C immersion temperature (courtesy Carl Hanser Verlag)

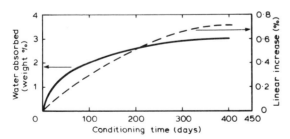

Figure 4.37  Water absorption and linear increase vs conditioning time for injection-moulded nylon 6 specimens, 120 mm × 20 mm × 4 mm, in air at 65% RH, 20°C (courtesy Carl Hanser Verlag)

Further observations have shown that the more crystalline the polyamide the less swelling is observed on absorbing moisture, and in general (due to differences in crystallinity and non-uniform moisture distribution in the component section) it is impossible to predict with accuracy the dimensional changes that can occur on exposure of any particular polyamide type to known environments.

A further factor contributing to this uncertainty is the presence of residual stresses in fabricated parts, particularly extrusions or mouldings. As moisture is absorbed these stresses are gradually relieved; particularly along the direction of moulding the linear increase due to moisture absorption is often found to be less than would be expected, since the relief of stresses usually results in some shrinkage. If required, a pre-annealing treatment may be used to

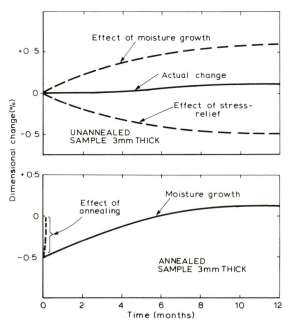

Figure 4.38 Typical dimensional changes of moulded nylon 66 in air (20% RH, 23°C) due to combined stress-relief and moisture absorption (courtesy Du Pont de Nemours)

relieve the internal stresses in the part prior to a moisturising treatment intended to prepare the part for service. Alternatively, moisturising can be used alone to anneal and prepare the part in

## Properties of polyamides

cases where the service temperature does not exceed the conditioning temperature. *Figure 4.38*[16] shows the effect of moisture absorption on dimensions of annealed and unannealed nylon 66 sheet.

Since moisture absorption may depend on many factors that are not always recognisable, it is usually necessary to carry out investigations to establish the dimensional change in individual cases. In service, changes in temperature may be more influential in effecting dimensional changes than moisture absorption. In many cases both factors play a part.

### 4.3.2 EFFECT OF MOISTURE

As discussed in section 4.3.1.2, polyamides can absorb small or large amounts of moisture according to type. Those types with a low $CH_2/CONH$ ratio (such as nylon 66 or nylon 6) can absorb over 9 per cent of moisture, and the consequent effect on the mechanical properties can be profound. The moisture is not necessarily absorbed to saturation level and there often exists a moisture gradient across the section at right angles to the exposed surface, which results in a corresponding gradation of properties. This is the usual result of moisture-conditioning (see Chapter 5), which among other reasons is carried out to bring the exposed polyamide surfaces into near moisture-equilibrium with the surroundings. However, since absorption and desorption of moisture in polyamides is a reversible process, a polyamide article initially brought to moisture-equilibrium with the environment may suffer an unacceptable change in mechanical properties due to moisture-content change, unless the environmental conditions are controlled. The moisture in polyamides usually acts like a plasticiser and its presence facilitates molecular-chain movement. This decreases stiffness and increases flexibility. Tensile and other moduli are reduced and elongation increased, therefore, with increasing moisture content.

#### 4.3.2.1 Effect on short-term properties

The effect of varying moisture content in nylon 6 on the short-term stress elongation curve is shown in *Figure 4.39*[46]. This shows the disappearance of a well-defined yield point at a moisture content around 1.2 per cent. *Figure 4.7*[16] shows the effect of moisture content on the short-term tension and compression stress–strain curve for nylon 66. The disappearance of the yield point is also shown to occur at a moisture content below 2.5 per cent.

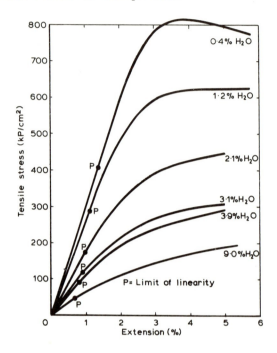

Figure 4.39 Stress–extension curves for nylon 6 at various moisture contents, 20°C (courtesy Carl Hanser Verlag)

*Figure 4.40*[47] shows that the short-term flexural yield stress of nylon 6 between $-40°C$ and $50°C$ decreases with increasing moisture content. The effect of moisture content on the notched Izod impact values of nylon 66 is most marked. *Figure 4.41*[16] shows that this value increases five-fold for a moisture increase of from 1 to 4 per cent. The indentation hardness of polyamides is very susceptible to change due to change in moisture content of the material since the property is measured at or near the surface, at which location the effect of gain or loss of moisture is most marked. Except under equilibrium conditions the indentation hardness is not necessarily related to the bulk moisture content. *Figure 4.8*, showing the relation between

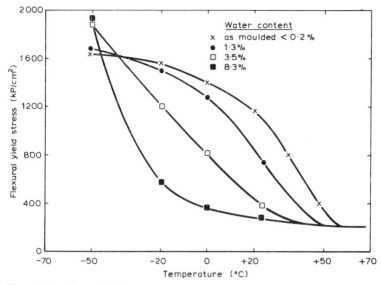

Figure 4.40  Flexural yield stress vs temperature for moulded nylon 6 at various moisture contents (courtesy Carl Hanser Verlag)

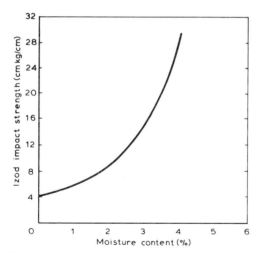

Figure 4.41  Effect of moisture content of nylon 66 on Izod impact strength at room temperature (courtesy Du Pont de Nemours)

indentation hardness and moisture content for a moisturised nylon 66 extrusion, illustrates this effect.

#### 4.3.2.2 Effect on long-term properties

Since the effect of moisture absorption of a polyamide is to decrease stiffness and increase flexibility, it follows that the creep resistance is also reduced. This is illustrated in *Figures 4.42*[18] and *4.21*[18],

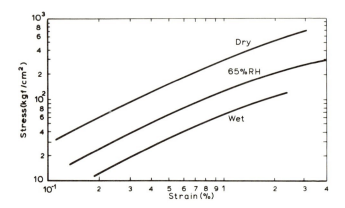

Figure 4.42 Isochronous stress–strain curves for nylon 66 at 20°C, 100 seconds, showing effect of moisture content (courtesy ICI Ltd)

referring to the isochronous stress–strain curves for, respectively, unfilled nylon 66 at 20°C and 33 per cent glass-filled nylon 66 at 20°C. The lower creep modulus resulting from moisture absorption implies also that the stress relaxation of a moist polyamide will be greater than that of the dry material. The long-term strength of the polyamide is also reduced by moisture absorbed. If stressed in tension the onset of necking occurs sooner in the moist material or at a lower stress, and the material tends to fail in a ductile rather than a brittle fashion. Dynamic fatigue failure occurs more readily with moist polyamide than with the dry material, due primarily to the higher damping capacity of the former. At higher frequencies, particularly, the greater amount of internally-generated heat due to absorption of energy by damping leads to earlier fatigue failure in the case of moist polyamide. *Figure 4.43*[16] illustrates this effect, showing flexural fatigue endurance limits (stress level at which test parts can

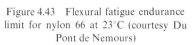
Figure 4.43 Flexural fatigue endurance limit for nylon 66 at 23°C (courtesy Du Pont de Nemours)
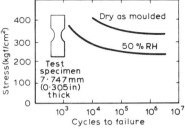

undergo $10^6$ cycles without failure) for nylon 66 both dry and in equilibrium at 50 per cent RH.

### 4.3.2.3 Effect on friction, wear, and bearing properties

It has already been stated in section 4.2.4.1 that the coefficient of friction of a polyamide surface increases with moisture content of the material. In conditions of continuous operation, such as in a bearing, the dynamic friction coefficient may not be permanent since frictional heat developed under continuous sliding (particularly at fast speeds) may reduce the moisture content in the interfacial layers to a near-dry state.

It is difficult to predict with any certainty the effect of moisture in a polyamide on its wear properties, because the wear rate is affected by a large number of factors that could be adversely or favourably influenced by moisture; since there has been little published work dealing specifically with the effect of moisture content on wear properties for thermoplastics, it would be unsafe to generalise.

In the very important practical field of unlubricated plastic bearings, the properties that affect the wear of the bearing plastic and the metal shaft have been discussed by Lancaster[48]. From his analysis it is possible to predict tentatively the effect of moisture in the bearing on its wear performance. Thus for abrasive wear (which occurs on metal shafts with a rough surface finish, and during the running-in phase of a plastic bearing) it should be expected that the wear rate decreases with increasing moisture content of the bearing, since the material at the interface becomes more plastic and the increased softness tends to decrease wear by the abrasive surface. After running-in, or for initially smooth surfaces, the fatigue type of wear contributes most to the total wear, the transition from the abrasive to the fatigue regime being gradual. Increase in moisture on the polyamide wear surface is likely to increase fatigue wear. Adhesive wear (which occurs with very smooth surfaces and is the result of transfer of plastic material to the metal shaft) is likely to be decreased with increase in moisture content, due to the reduced adhesive capacity of the transfer film. Prediction here is most uncertain, however, because of the many factors, such as the stability and performance of the transfer film, influencing this type of wear.

Fillers (for example, graphite or $MoS_2$) are known to reduce the wear of dry polyamide bearings but may be adversely affected by moisture due to inhibition of the stability of the transfer film. This can result in irregular or irreproducible wear behaviour.

For bearings made from polyamides of high potential moisture uptake, wear rate is dependent on both the moisture content of the bearing and the environmental conditions. For designing such bearings it is recommended that simulated service tests be carried out using both the polyamide material and the shaft material in the state and in the environment likely to be encountered in service.

## 4.4 Physical properties

The physical properties of polyamide plastics are of considerable interest to the convertor who has to turn granular or powdered polymer, obtained from a raw material supplier, into useful shapes, to the designer who has to take account of property data in his calculations, and to the end-user who expects a given performance from an article or component from a knowledge of its listed properties.

Basic properties data, where the values have been determined using standardised test equipment, are readily available from suppliers of polyamide resins and are widely quoted in material specifications adopted by national or international authorities or services organisations acting as end-users. A considerable amount of useful design data on physical properties is also available from the same source. In addition, an important advance towards filling a long-felt need to present properties of polyamides and plastics in general in a form suitable for designers has been the publication of BS 4618[49], which gives guidelines to achieve this aim.

The remainder of this chapter is devoted to a general discussion of the properties of commercial polyamides under various headings. Where relevant, properties will be interpreted in terms of the structural concepts outlined in section 4.1.1.

## Properties of polyamides

### 4.4.1 APPEARANCE

Uncompounded polyamides are generally transparent and nearly colourless in the molten state and opaque white or yellowish-white in the solid, partially crystalline state. Those polymers possessing a low degree of crystallinity tend to become transparent and some copolymers are completely transparent. Impurities originating from the raw materials can impart a yellowish coloration to polyamides, as can degradation products arising from reactions during the polymerisation process. Surface oxidation, which is relatively rapid above 120°C, can show up as a yellow or brown coloration on the surface of granular moulding powder and other forms. Unless oxidation has been extensive, this is usually confined to a shallow layer at the surface.

The crystalline polyamides have a shiny surface when first formed from the melt. Surfaces are smooth and the low coefficient of friction of the solid imparts good flow and feed properties to granular moulding powders.

The crystalline polymers can be compounded with a large number of materials for decorative or functional uses. These may modify the original coloration or (as in the case, for example, of heat or light stabilisers) prevent further discoloration. Complete modification is possible by using self-colouring dyes or pigments dispersed evenly throughout the polymer or by dyeing the surface of the fully polymerised material in a separate operation. Processes for metal plating by electrolytic chemical deposition or metal spray processes are also known.

### 4.4.2 SPECIFIC GRAVITY

The specific gravity of a polyamide is dependent on its type and degree of crystallinity, the room-temperature values of all types covering the approximate range 1.01 to 1.16. For a given type specific gravity increases with degree of crystallinity, the former increasing approximately 0.01 units per 10 per cent increase in the latter. Other things being equal, increasing the number of methylene groups in the chain separating repeating functional groups leads to decreasing specific gravity. This is illustrated in *Table 4.20*, which shows approximate ranges for commercial polyamides.

In processing, the rate of cooling of a polyamide from the melt, in so far as the crystallinity is affected, determines the final specific

## Properties of polyamides

**Table 4.20** SPECIFIC GRAVITY RANGES FOR COMMERCIAL POLYAMIDES (ROOM-TEMPERATURE VALUES)

| Polyamide | SG range |
|---|---|
| 66 | 1.13–1.16 |
| 6 | 1.12–1.15 |
| 8 | 1.06–1.08 |
| 6.10 | 1.07–1.09 |
| 11 | 1.03–1.05 |
| 12 | 1.01–1.04 |

gravity of the product. Where cooling rate varies across the section, as in extrusion, specific gravity also varies, and this can give rise to a degree of inhomogeneity in the material. Drawing and other forming processes induce orientation and increase the tendency to crystallise, producing a corresponding increase in specific gravity. Since the specific gravity of polymers can be measured rapidly and accurately by a variety of techniques, notably the density gradient column method, this property is commonly determined for quality testing in conversion processing of polyamides and in fault-finding investigations.

### 4.4.3 THERMAL PROPERTIES

The thermal properties of polyamides may be conveniently considered from several viewpoints. For example, changes occurring at fixed points, either first- or second-order transition points, may be observed when a polyamide is heated from ambient temperature to melting or, conversely, cooled from the melt to ambient temperature. Again, the absorption or evolution of heat at any temperature or at the transition points (defining, respectively, the specific heat and latent heat of transition) may be observed. The rate of transfer of heat at any temperature is quantified by the thermal conductivity, while the volume change in the polyamide per unit temperature change is defined by the thermal expansion. Allied to the transition points (and of more significance perhaps to the engineer) are heat deflection temperatures at which, under uniform temperature rise and fixed arbitrary loads, deformation of the material by an arbitrary fixed amount occurs.

### 4.4.3.1 Transition points

It is well known that all crystalline solids have sharp melting points while amorphous solids have melting or softening ranges. As would be expected, partially crystalline polymers such as polyamides have melting ranges that are wide or narrow according to the degree of crystallinity they possess. On the whole the homopolyamides such as nylon 6 and nylon 66 have narrow melting ranges while the copolyamides have much wider melting ranges. The melting point may be considered thermodynamically as a first-order transition involving a change of state, i.e. solid to liquid. The amorphous content of a polyamide is, however, not subject to first-order transitions. Just below the crystalline melting point the amorphous segments of the molecular chain undergo large-scale molecular motion characteristic of the rubbery state. These motions become progressively smaller as the temperature is reduced until the glass-transition temperature, $T_g$, is reached, when the long-range molecular movement is replaced by rather more local movement of atoms or groups of atoms moving against local restraints. The occurrence of the glass transition and first-order transitions in polyamides must be fully appreciated to enable an explanation to be made of the phenomena that occur on traversing wide temperature ranges. The existence of both crystalline and amorphous fractions in homopolyamides causes melting over a range of temperature.

The mechanical properties of polyamides obviously change in passing through the glass transition, the magnitude of the effect varying according to the degree of crystallinity of the polymer. The general features are shown in *Table 4.21* from Billmeyer[50]. The intermediate crystallinity range shown in the table corresponds to the large majority of polyamides of interest in engineering applications.

Published data on glass transitions vary widely since there is a variety of methods of detecting the change, some depending on bulk properties (e.g. expansion coefficients), others on methods related to molecular motion (e.g. dielectric loss measurement). *Table 4.22* gives the glass-transition temperature, $T_g$, for some common linear polyamides, together with the crystalline melting point, $T_m$.

The melting point, $T_m$, of copolyamides depends on both composition and the susceptibility of the mixed components to form either isomorphous or anisomorphous crystals. These types have already been discussed at length, and the melting point vs composition curves are shown in *Figure 3.5*.

## Properties of polyamides

**Table 4.21** PROPERTIES OF CRYSTALLINE POLYMERS—MAIN PROPERTY FEATURES (courtesy John Wiley)

| Temperature range | Degree of crystallinity | | |
|---|---|---|---|
| | Low (5–10) | Intermediate (20–60) | High (70–90) |
| Above $T_g$ | Rubbery | Leathery, tough | Stiff, hard, brittle |
| Below $T_g$ | Glassy, brittle | Hornlike, tough | Stiff, hard, brittle |

**Table 4.22** TRANSITION TEMPERATURE FOR LINEAR POLYAMIDES

| Type | $T_g$ (°C) | $T_m$ (°C) |
|---|---|---|
| 4 | 75 | 265 |
| 6 | 50 | 228 |
| 8 | 51 | 209 |
| 10 | 43 | 192 |
| 11 | 46 | 194 |
| 66 | 57 | 265 |
| 6.10 | 50 | 228 |
| 6T (terephthalic acid) | 180 | — |

### 4.4.3.2 Thermal expansion

The thermal expansion of polyamides depends on the stability of the crystal structure of the type under consideration and on the degree of crystallinity. Where the structure is particularly stable (as in nylon 66) the expansion is lower than, for example, for nylon 6.10 where the stability is less. The longer the alkyl segment between functional groups in the chain the less are the intermolecular forces and the greater is the expansion coefficient. By the same token the amorphous fraction of a polyamide has a greater thermal expansion than the crystalline fraction. The above generalisations are illustrated in *Table 4.23*[51], which shows the linear expansion coefficients of linear homopolyamides of different degrees of crystallinity.

Addition of inert fillers to polyamides lowers the thermal expansion. For example unfilled nylon 66 has a linear expansion coefficient of $9 \times 10^{-5}$ per °C. When filled with finely divided graphite the coefficient can be reduced to $7.6 \times 10^{-5}$ per °C. Fibrous fillers

## Properties of polyamides

have a similar effect in reducing thermal expansion and in addition may have a special effect due to fibre orientation. These effects for glass-fibre filler are shown in *Table 4.24*[52].

**Table 4.23** LINEAR EXPANSION COEFFICIENT OF POLYAMIDES IN RANGE 0–60°C: EFFECT OF CRYSTALLINITY (courtesy Carl Hanser Verlag)

| Polyamide | Percentage crystallinity | Linear expansion coefficient (per °C × 10⁻⁵) |
|---|---|---|
| 6 | 35 | 6.5 |
|   | 15 | 10.0 |
| 66 | 35 | 7.0 |
|    | 20 | 9.0 |
| 6.10 | Quenched | 13.0 |
|      | Slow-cooled | 8.0 |
| 8 | Quenched | 12.0 |
|   | Slow-cooled | 10.0 |
| 11 | Quenched | 12.5 |
|    | Slow-cooled | 11.0 |

**Table 4.24** EFFECT OF GLASS-FIBRE FILLER AND FIBRE ORIENTATION ON LINEAR EXPANSION COEFFICIENT OF NYLONS (courtesy ICI Ltd)

|  | Unfilled nylon 66 | 33% glass-fibre filled nylon 66 | |
|---|---|---|---|
|  |  | Parallel to moulding axis | Randomly oriented |
| Linear coefficient of thermal expansion (per °C × 10⁻⁵) | 9 | 2.8 | 5 |

It is usually found that the expansion measured in the direction of fibre orientation is less than that found if the fibres are randomly oriented. This feature must be taken into account when designing with fibre-filled polyamides.

The thermal expansion of a polyamide can also be modified by including in the formulation small amounts of special inorganic compounds. For example addition to nylon 66 of about two per cent of molybdenum disulphide results in a 40 per cent reduction in linear thermal expansion. This is attributed to the refinement in crystal structure brought about by the additive.

### 4.4.3.3 Thermal conductivity/specific heat

For unfilled polyamides the range of thermal conductivity values is quite narrow, the type and structure of the polymer having relatively little influence. The effect of temperature on conductivity is also slight. For example, nylon 6 is reported to drop in thermal conductivity by only 16 per cent when the temperature is raised from 20°C to 100°C[53]. As expected both crystalline and oriented regions of a polyamide show greater conductivity than the amorphous material of the same composition. Typical thermal conductivity values for common linear polyamides are shown in *Table 4.25*[54].

**Table 4.25** THERMAL CONDUCTIVITY OF POLYAMIDES (courtesy Carl Hanser Verlag)

| Polyamide type | 6 | 6.6 | 6.10 | 8 | 11 | 12 |
|---|---|---|---|---|---|---|
| Thermal conductivity (kcal/m h °C) | 0.24 | 0.21 | 0.19 | 0.20 | 0.25 | 0.28 |

The values given in the table do not include the effects of factors conferred on the polymer in processing, such as built-in stresses, residual moisture or monomer, which may affect the conductivity. These same factors also modify the thermal expansion. Designers therefore should try as far as possible to obtain data referring to the material in its manufactured or supplied condition. For example, thermal conductivity should be known for the material in directions parallel to and across the direction of melt flow as well as for fully annealed material. Fillers modify the thermal conductivity according to their type and loading in the polymer, and the general effect can be forecast from knowledge of the filler thermal properties.

Specific heats of the nylons quoted in *Table 4.25* vary from 0.4 to 0.58 cal/g °C. The values for filled grades are lower.

### 4.4.3.4 Deflection temperature under load

The retention of mechanical properties with rise in temperature is a highly desirable feature in plastics applications and the majority of polyamides possess this property to a marked degree. One group of methods of assessing the short-term temperature effect on mechanical properties is based on measuring the deflection temperature under load. Measurement can be made using several methods detailed in well-known standards, such as the Deflection Temperature under

## Properties of polyamides

Load of ASTM D 648, the Martens test of DIN 53458 and the Vicat test of VDE 0302/III, both the latter being German national standards.

While the data obtained by these methods cannot be used directly in design calculations they are useful in the selection of materials. Of the three methods mentioned, the deflection-temperature method of ASTM D 648 is to be preferred since the loadings used relate more to practical situations than do those of the other methods, and the values can be accepted with more confidence by designers. Deflection temperature by the ASTM method also shows up markedly the stiffening effect of fillers in the polyamide. *Table 4.26*[18] shows the effect of nylon type and glass-fibre filler on deflection temperature. The marked increase in deflection temperature of nylon 66 and nylon 6 on addition of glass-fibre should be noted. For this reason the filled grades are preferred in elevated-temperature applications where rigidity is required.

**Table 4.26** DEFLECTION TEMPERATURE UNDER LOAD FOR NYLON TYPES: METHOD AS ASTM D 648 (courtesy ICI Ltd)

|  |  |  |  |  | Glass-filled 33% | |
|---|---|---|---|---|---|---|
|  | 66 | 6 | 6.10 | 11 | 66 | 6 |
| Deflection temperature (°C) |  |  |  |  |  |  |
| at 4.6 kgf/cm$^2$ | 190 | 155 | 150 | 145 | 254 | 216 |
| at 18.5 kgf/cm$^2$ | 75 | 70 | 57 | 50 | 245 | 194 |

### 4.4.4 ELECTRICAL PROPERTIES

The electrical properties of nylons, although inferior to those of some other thermoplastics (notably polyolefins and polystyrene), are quite adequate in low-frequency applications. The choice of nylons for electrical service is often determined by their marked superiority in mechanical properties.

Almost all the electrical properties of polyamides are strongly affected by the presence of moisture in the polymer, and the capacity of a particular polyamide to absorb moisture can affect its suitability for a particular electrical application. Other factors that influence electrical properties are temperature, frequency of applied voltage, degree of crystallinity, $CH_2/CONH$ ratio in the molecule, and thickness of section. The general effect of all these factors is summarised in *Table 4.27*.

**Table 4.27** EFFECT OF FEATURE CHANGES ON ELECTRICAL PROPERTIES OF POLYAMIDES

| Increase in: | Volume/surface resistivity | Dielectric loss (tan δ) | Dielectric constant | Dielectric strength |
|---|---|---|---|---|
| Moisture content | − | + | + | − |
| Temperature | − | +* | +* | − |
| Applied voltage frequency |  | + | − |  |
| Degree of crystallinity | − | + | + |  |
| $CH_2$/CONH ratio | + | − | − | + |
| Thickness of section |  |  |  | −‡ |

+ Increase in property
− Decrease in property
* Can decrease above a point on the temperature loss curve
‡ For thin sections

### 4.4.4.1 Resistivity

In very dry conditions polyamides at normal temperature have volume resistivities between $10^{14}$ and $10^{15}$ ohm-cm. Resistivity decreases with increasing moisture content. For example, resistivity of moisture-saturated nylon 6 is $10^8$–$10^9$ ohm-cm while that of saturated nylon 11 is $10^{12}$ ohm-cm; the difference is due largely to the difference in saturated moisture contents, which are for nylon 6 and nylon 11 about 10 and 1 per cent respectively.

The volume resistivity of polyamides is strongly dependent on temperature. For example, for dry nylon 66, the value reduces by about one order per 15°C rise in temperature. The reduction is less for moist nylon 66. Surface resistivity of polyamides is strongly dependent on moisture, and even a short-time exposure of the dry material to normal ambient humidity will substantially decrease the value of this property. Since most polyamides have an appreciable moisture uptake, compared with other thermoplastics, resistivities tend to be low. This to some extent prevents accumulation of static charge on the surface, which property is useful in applications where these charges constitute a fire or explosive risk. *Table 4.28*[55] shows the combined effect on volume resistivity of polyamides of different types and moisture contents.

## Properties of polyamides

**Table 4.28** EFFECT OF TYPE AND MOISTURE CONTENT OF POLY-AMIDES ON VOLUME RESISTIVITY (courtesy Carl Hanser Verlag)

| Polyamide type | Moisture condition | Volume resistivity (ohm-cm) |
|---|---|---|
| Nylon 6 | Dry | $3-5 \times 10^{14}$ |
| | 24 h water-exposed | $6 \times 10^8$ |
| | Moisture-saturated | $3 \times 10^8$ |
| Nylon 66 | Dry | $8 \times 10^{14}$ |
| | 24 h water-exposed | $3.5 \times 10^{11}$ |
| | Moisture-saturated | $3 \times 10^8$ |
| Nylon 6.10 | Dry | $3 \times 10^{14}$ |
| | 24 h water-exposed | $6 \times 10^{11}$ |
| | Moisture-saturated | $2.5 \times 10^{10}$ |
| Nylon 11 | Dry | $4 \times 10^{14}$ |
| | Moisture-saturated | $1 \times 10^{12}$ |
| Nylon 12 | Dry | $5 \times 10^{14}$ |
| | Moisture-saturated | $8 \times 10^{12}$ |

### 4.4.4.2 Permittivity and dielectric loss

Like resistivity, permittivity $\varepsilon$ and loss tangent tan $\delta$ are both strongly influenced by moisture in the polymer, the values of each property increasing with moisture. The effect, as shown in *Figure 4.44*[56], is considerably less for nylon 11, 12 or 6.10 than for nylon 66 or 6.

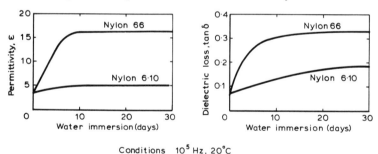

Figure 4.44 Effect of moisture on permittivity and dielectric loss of nylons (courtesy Carl Hanser Verlag)

Dielectric loss generally increases as temperature is raised above ambient, the crystalline material showing a higher loss than the amorphous. For some polyamides, such as nylon 6 and 66, the temperature-loss curve shows a maximum. In the case of nylon 6 this occurs between 60°C and 100°C and corresponds to an inflection in the temperature–mechanical shear modulus curve. This pheno-

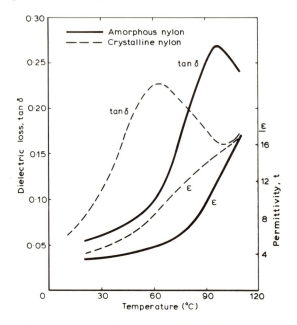

Figure 4.45  Dielectric loss/permittivity vs temperature for crystalline and amorphous dry nylon 6 at $10^5$ Hz (courtesy Carl Hanser Verlag)

menon is put down to structural changes in the molecule, and indicates the onset of the softening range. The changes are shown in *Figure 4.45*[57].

### 4.4.4.3  Dielectric strength

The dielectric strength of polyamides, like specific resistance but to a much smaller degree, decreases with increase in both moisture and temperature of the polymer. The change becomes less marked as the $CH_2/CONH$ ratio in the molecule increases. In common with other plastics insulating materials, polyamide dielectric strength (as determined by the short-term test of ASTM D 149) increases with decrease in the thickness of the section under test. *Figure 4.46*[16],

## Properties of polyamides

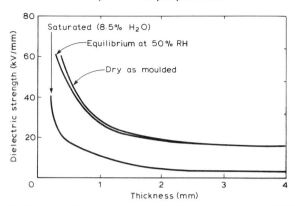

Figure 4.46 Short-time dielectric strength vs thickness for nylon 66 at 23°C (courtesy Du Pont de Nemours)

showing the variation of dielectric strength with thickness for nylon 66 at two moisture levels, illustrates the trend. Such relationships are seldom more than a guide to designers since the breakdown voltage depends also on the duration of the applied stress. Testing under simulated service conditions is, therefore, recommended.

### 4.4.4.4 Tracking resistance

There is little published data on the tracking characteristics of polyamides and only a few reliable methods for measuring this property are known. Tests carried out by the Electrical Research Association using the Comparative Tracking Index (CTI) method detailed in BS 3781 shows the rating for nylons to be high, indicating a low tracking susceptibility compared with a number of other commercial engineering plastics. Values are shown in *Table 4.29*.

### 4.4.4.5 Effect of frequency

Polyamides, especially those with low $CH_2/CONH$ ratio, are not outstanding in performance in high-frequency applications. For unfilled polyamides permittivity decreases with frequency, the effect being most marked as the moisture content of the material increases, but for the majority of applications at a frequency of 50 Hz the loss

## Properties of polyamides

**Table 4.29** COMPARATIVE TRACKING INDEX OF COMMERCIAL ENGINEERING PLASTICS

| Material | CTI (BS 3781) |
|---|---|
| PTFE | Over 700 |
| Nylon 66 | Over 700 |
| Nylon 6.10 | Over 700 |
| Polyethylene terephthalate | Over 700 |
| Polyacetal copolymer | Over 700 |
| Polypropylene | Over 700 |
| Glass-fibre filled nylon | Less than 400 |
| Rigid PVC | 325 |
| Glass-fibre filled polyester | 250 |
| Polystyrene | 250 |
| Polycarbonate | 180 |

is not substantial. Dielectric loss in low-moisture-content unfilled polyamides can also increase at low frequency and decrease at high frequency. With polyamides of high moisture content this trend may be reversed. The effect is shown in *Figure 4.47*[18]. Glass-fibre filled nylon exhibits a more or less similar effect but here absorption of

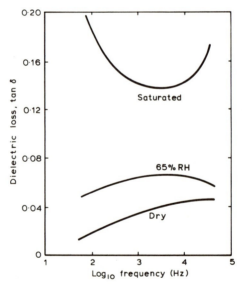

Figure 4.47  Loss tangent vs frequency for unfilled nylon 66 at 23°C, showing effect of moisture (courtesy ICI Ltd)

## 4.4.4.6 Effect of fillers

Reinforcing particulate or fibrous mineral fillers such as glass, which are intended to improve the mechanical properties of polyamides, have relatively little effect on electrical properties in the dry state. In the moist condition permittivity and dielectric loss tend to be higher for the filled grades than for the corresponding unfilled grade. Fibrous fillers usually orient themselves in processing, and it has been found that the electrical insulation properties measured across the direction of orientation are superior to those measured at right angles to this direction. Carbon fibres are a special case since the fibres are themselves electrical conductors, and the electrical properties of polyamide carbon-fibre composites depend on total fibre content, subdivision of the fibres, and orientation of the fibres.

### REFERENCES

1. Staudinger, H. and Heuer, W., *Berichte*, **63**, 222 (1930)
2. Allen, P. W. (ed.), *Techniques of polymer characterisation*, Academic Press, New York (1959)
3. Price, F. P., *J. Am. chem. Soc.*, **74**, 311 (1952)
4. Khoury, F., *J. Polymer Sci.*, **33**, 389 (1958)
5. Vieweg, R. and Muller, A., *Kunststoff Handbuch*, Vol. VI, *Polyamide*, 480, Carl Hanser Verlag, Munich (1966)
6. Gabler, R., *Kunststoffe-Plastics 3*, 5 (1956)
7. Ackhammer, B. G., Reinhard, F. W. and Kline, G. M., *J. Res. National Bureau of Standards*, **46**, 391 (1951); *J. Applied Chem.*, **1**, 301 (1951)
8. Strauss, S. and Wall, L. A., *J. Res. National Bureau of Standards*, **60**, 39 (1958)
9. Kamerbeek, B., Kroes, G. H. and Grolle, W., *Thermal degradation of some polyamides*, from Society of Chemical Industry Monograph No. 13 (1961)
10. Harding, G. W. and MacNulty, B. J., *The embrittlement of polyamides*, from Society of Chemical Industry Monograph No. 13 (1961)
11. Moore, R. F., *Polymer*, **4**, 493 (1963)
12. Kroes, G. H., *Rec. Trav. Chim.*, **82**, 979 (1963)
13. Chottiner, J. and Bowden, E. B., *Materials in design engineering*, 97 (1965)
14. Charlesby, A., *Plastics*, **18**, 70, 142 (1953)
15. Sisman, O. and Bopp, D. C., Oak Ridge National Laboratory, ORNL 928, 85 (1951)
16. Du Pont de Nemours, *Zytel nylon resin design handbook*
17. Ritchie, P. D. (ed.), *Physics of plastics*, Plastics Institute Monograph, Newnes–Butterworths (1965)
18. Imperial Chemical Industries Ltd, *Maranyl data for design*, Plastics Division, Technical Service Note N 110
19. Ref. 5, 519

20  Ref. 5, 522
21  Sandhof, U., *Kunststoffe*, **61**, 777 (1971)
22  Ref. 5, 523
23  Horvath, L., *Plastics*, **33**, 415 (1968)
24  Dorst, H. G. and Fischer, K., *Plastics*, **33**, 1290 (1968)
25  Gotham, K. V., *Plastics and Polymers*, **37**, 309 (1969)
26  Riddell, M. N., Koo, G. P. and O'Toole, J. L., *Polymer Engng and Sci.*, **6**, 363 (1966)
27  Oberbach, K., *Kunststoffe*, **63**, 35 (1973)
28  Dally, J. W. and Carillo, D. H., *Polymer Engng and Sci.*, **9**, 434 (1969)
29  Wiegand, H. and Vetter, H., *Kunststoffe*, **56**, 76 (1966)
30  Shooter, K. V., *Proc. R. Soc.*, **212A**, 448 (1952)
31  Bowden, F. P. and Tabor, D., *The friction and lubrication of solids*, Part 1, Clarendon Press, Oxford (1954)
32  Polypenco Ltd, private communication
33  Lancaster, J. K., *Wear*, **14**, 223 (1969)
34  Reddy, D. and Nain, B. S., *Wear testing*, British Hydromechanics Research Association Report TN 940 (1968)
35  Pratt, W., *Lubrication and wear*, 404, Elsevier (1967)
36  O'Rourke, J. T., *Proceedings of first, second and third fluorocarbon design conferences*, SPI Inc., New York (1966)
37  Polypenco Ltd, *Bearing design*, technical brochure
38  Hussey, E. H., Polypenco Ltd, private communication
39  Heinz Faigle, Austria, *PAS material*, technical brochure
40  Ref. 5, 431
41  Ref. 5, 470
42  Ref. 5, 464
43  Ref. 5, 472
44  Ref. 5, 465, Fig. 18
45  Ref. 5, 465, Fig. 19
46  Ref. 5, 509
47  Ref. 5, 532
48  Lancaster, J. K., *Plastics in bearings*, Plastics Institute Conference (Feb. 1973)
49  British Standards Institution, BS 4618, *Recommendations for the presentation of plastics design data* (1970–1974)
50  Billmeyer Jr., F. W., *Textbook of polymer science*, 206, John Wiley (1971)
51  Ref. 5, 547
52  Imperial Chemical Industries Ltd, *Maranyl nylon*, Plastics Division Technical Service Note N 104
53  Knappe, W., *Plaste und Kautschuk*, **4**, 189 (1962)
54  Ref. 5, 551
55  Ref. 5, 554
56  Ref. 5, 557
57  Ref. 5, 558

# 5
# Conversion processes

## 5.1 General

The production of nylon polymer from raw material and intermediates, described in Chapter 3, is the first step in a sequence of processes that ends with the production of nylon in the form suited to the end-user's requirements. The conversion of nylon granules or 'chips' produced in the polymerisation plant to useful forms is probably, in terms of capital invested and labour employed, the largest section of the nylon industry. Expansion of the industry has been accelerated by the adaption of conventional types of conversion equipment used for other thermoplastics to enable the nylon types available to be processed.

This chapter details the main types of conversion processes used for the nylons and discusses how the properties of nylons as a class must be taken into account in the design and control of the equipment. The effect of individual processing methods on the bulk properties of the fabricated material is also discussed.

## 5.2 Injection moulding

Injection moulding is by far the most common and versatile method of converting nylon to useful shapes. It is particularly suited to the production of long runs of small components required in the mass-production of automobiles, domestic appliances and the like. The method itself has the advantage that, dependent on the skill of the mould designer, forms close to the final shape are produced in one operation and second operations can be kept to the minimum. Some mouldings, however, require some sort of finishing operation, and this is discussed in section 5.7. Engineering components of high reliability are now produced in vast numbers by injection moulding.

## 5.2.1 MOULDING CHARACTERISTICS OF NYLON

The injection-moulding characteristics of the commercial nylons differ from those of other mouldable materials in four main respects:

(1) low viscosity, high-temperature melt;
(2) narrow plastic range between melting and degradation;
(3) susceptibility to moisture;
(4) abrupt transition from solid to melt.

The consequence of the above features is discussed in detail in the following.

A low melt viscosity means a very fluid mix, and a special nozzle is required to overcome the problem of dribble or drooling. In its simplest form this can be a reverse-taper nozzle with the narrow end towards the cylinder. A more positive type is the spring-loaded valve nozzle normally sealed off but opening with the forward stroke of the ram. The mating parts of moulds and other surfaces must fit well to prevent leakage when the highly fluid mix is injected. Flash is difficult to remove since the nylons are tough and resilient.

Contraction of the highly fluid melt is high (16–17 per cent) due to the injection temperature being higher than for other thermoplastics, and also because of the crystallinity developed on cooling. To counteract this contraction a high follow-up pressure is required.

With nylons, pressure losses in moulding should be less than with most other thermoplastics due to the fluidity of the former. In practice, however, due to the sharp melting points of the nylons, the material tends to freeze in the runners if the temperature drops and generally higher pressures than used for other thermoplastics are required for filling the mould. In this respect the size of gate used is important; too narrow a gate tends to increase fluid friction, raise the temperature of the melt, and increase the risk of thermal degradation; too wide a gate, although reducing mould fill time, can also lead to 'jetting'.

The many other effects of processing a fluid mix on the properties of the final moulding are described in detail in an excellent monograph on the subject[1].

The narrow temperature range available for the processing of the majority of nylons means that the heating system of the injection-moulding machine must be capable of a high degree of control. For example a variation at the cylinder wall in excess of $\pm 1°C$ is considered unsatisfactory. In this connection it should be realised that the heat input required to bring nylons to moulding temperatures is one of the highest among common moulding thermoplastics, as can be seen from *Table 5.1*[1].

Table 5.1  ENTHALPY OF PLASTICISED MATERIAL

| Material | Approximate total heat at moulding temperature | |
|---|---|---|
| | kcal/kg | Btu/lb |
| General-purpose polystyrene | 65 | 120 |
| Low-density polyethylene | 140–170 | 250–300 |
| High-density polyethylene | 170–200 | 300–350 |
| Acetal | 100 | 180 |
| Nylon | 150–180 | 275–325 |
| Acrylic | 70 | 125 |
| Polypropylene | 140 | 250 |
| PVC | 40–85 | 70–150 |
| Cellulose acetate | 70 | 125 |
| Cellulose acetate butyrate | 65 | 120 |
| ABS | 75–95 | 140–170 |
| Styrene acrylonitrile | 65–85 | 120–150 |

Increase in temperature of a nylon beyond its permissible range for moulding, whether arising from loss of temperature control or from development of excess frictional heat due to passage of the melt through the gate, tends to increase degradation of the material and impairs its properties. Degradation is also increased if the dwell time of the plasticised nylon in the cylinder is excessive, as can occur when very large machines are used for small mouldings. Mismatching of moulding size with machine size should therefore be avoided. Weakness at weld lines due to confluence of the two melt fronts within the mould is increased by air oxidation in an excessively hot melt. Again, at high rates of filling the entrapment of air and its subsequent compression can cause overheating, the hot spots being revealed by brown or black markings on the moulding surface. This particular fault may be overcome by proper venting of the mould.

The influence of moisture on the bulk properties of the nylons has already been discussed in Chapter 4. In relation to injection moulding it is important that the moisture content of the moulding powder delivered in the feed hopper does not exceed about 0.2 per cent, otherwise the mouldings will show surface streaks or splash marks and moulding itself will be difficult. The powder is therefore supplied in sealed containers, which should not be opened until just before use. All unused powder in the moulding machine must be resealed (and possibly even redried before resealing) to reduce moisture content to below the permitted maximum. Experience has shown[2] that sprues and runners may be used as rework without prior drying provided they are used immediately after they are made, by feeding back into the machine, otherwise post-drying will be required. There may, however, be good reasons for avoiding large amounts of rework, for instance if a large reduction of molecular weight (detected by solution viscosity measurements) is shown to occur on moulding. Recycling of such material can lead to a deterioration in useful properties.

The linear nylons differ, according to type, in their rate of moisture uptake, nylon 6 of the commercial types having the highest uptake rate and nylon 11 and 12 the lowest.

## 5.2.2  TYPE CHARACTERISTICS

While all nylons possess in some degree the moulding characteristics described above, variations occur for individual types. These are discussed below.

### 5.2.2.1  Nylon 66

This type is the most popular for commercial production in the UK and US. The usable plastic range is about 265–300°C, between the melting point and temperature of appreciable degradation. The maximum temperature control is required for this type, and individual control on all the cylinder zones and the nozzle of the equipment is advised. Nylon 66 is very susceptible to moisture and the greatest care must be taken to ensure that it is kept dry prior to moulding. Due to the high melt temperature and crystallinity developed in the moulding the volumetric shrinkage (16–17 per cent) is among the highest of all the nylons.

### 5.2.2.2  Nylon 6

This type is more popular than 66 on the continent of Europe, and among polyamides accounts for the greatest of world moulded tonnage. Compared to nylon 66, nylon 6 has a lower melting point (220°C) and a wider plastic range (220–290°C). Equilibrium moisture

Conversion processes 149

content either at ambient atmosphere conditions or in water immersion is significantly higher than for nylon 66.

### 5.2.2.3 Nylon 6.10

This type has a similar melting point (222°C) to nylon 6, but a wider usable melting range for moulding. Mould shrinkage is less than for nylon 66.

### 5.2.2.4 Nylon 11 and nylon 12

The melting points of these types are, respectively, 186°C and 175°C, the lowest of the commercial nylons, but their injection moulding processing temperatures are generally between 220°C and 280°C and depend on the type of moulding machine. Mould shrinkage is probably the lowest of all commercial nylons, as is moisture absorption. These types are therefore the most dimensionally stable of the unfilled nylons. The granules for extrusion must however still be dried to lower than 0.2 per cent moisture content. For these nylons, particularly nylon 11, very many unfilled grades, covering a range of melt viscosities, and grades with a wide variety of fillers are available.

### 5.2.2.5 Other types

Combinations of the nylons mentioned above are available in the form of copolymers, for injection moulding. These materials are generally of low crystallinity and low melting suitable for components requiring impact resistance, transparency and other special properties.

### 5.2.3 MOULDING MATERIALS

There is available on the market today a vast range of nylon moulding material types and grades. These are compounded primarily to give a final product with prescribed properties but additionally, in some cases, to improve processability.

Processability for any one nylon type is largely dependent on the molecular weight and distribution of the base polymers. For example the melt flow characteristics can be altered by changing average molecular weight to suit the type of moulding being produced. To impart special properties to the moulding, additives are used in the powder. For instance stabilisers may be added to improve heat stability, light stability or susceptibility to hydrolysis. Nucleating agents such as colloidal silica are used to improve the texture and increase the degree of crystallinity of the product and the rate of crystallisation of the nylon melt, at the same time reducing thermal expansion and moulding cycle time. Other additives may be used to improve fire-retardancy, lubricity and wear resistance. These additives are generally used in small percentages and the basic mechanical properties are not substantially affected.

On the other hand filled grades (where the filler, usually in fibre form, is added in substantial amounts to improve the mechanical properties of the moulding) have now found widespread use in nearly every field of application. Nearly all filled grades use short glass-fibres as reinforcement; this can be incorporated in proportions exceeding 40 per cent by weight of the compound. The fibres in these powders are homogeneously dispersed in the polymer matrix. The mouldings, compared with those produced from unfilled nylon, exhibit marked increase in tensile strength, rigidity, and deflection temperature under load, all due to the presence of the reinforcing filler. At the same time the dimensional stability of the moulding is significantly improved.

Carbon-fibre-filled nylon moulding materials have recently become commercially available. This fibre confers greater stiffness on the moulding than glass fibre at equivalent volume loadings, allowing weight savings when used in stressed parts. The use of this material is thus favoured in the aerospace industry.

## 5.2.4 MOULDING MACHINES

The basic design of a typical injection-moulding machine for thermoplastics is shown in *Figure 5.1*[4]; *Figure 5.2*[3] illustrates a typical injection-moulding time cycle. To achieve higher rates of plasticisation and obtain better mixing and pressure control the basic design has been improved, and the single-cylinder single-screw preplasticising machine illustrated in *Figure 5.3*[4] is now the preferred type for moulding nylon.

Injection-moulding machines are usually rated in terms of shot capacity and plasticising capacity. The former is the volume of the

## Conversion processes

material that can be delivered in one injection cycle, but it is often expressed as the weight in grams of general purpose polystyrene that can be delivered in one cycle. The shot capacity of nylon, which is of greater density than polystyrene, is usually less than calculated on this basis. Preplasticising raises the value for nylon to the

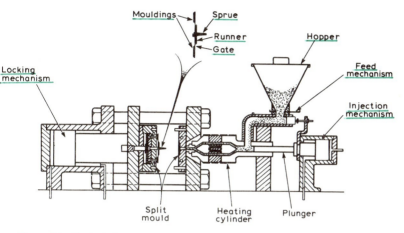

Figure 5.1  Typical plunger-type injection-moulding machine (courtesy ICI Ltd)

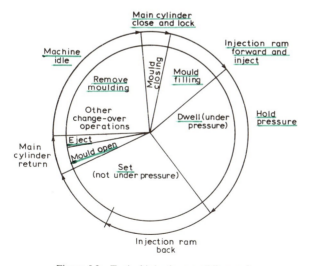

Figure 5.2  Typical injection-moulding cycle

theoretical figure or even in excess of it. Plasticising capacity is the rate (usually in kg/h or lb/h) at which the machine is able to plasticise the moulding powder. For a given machine this capacity depends largely on the enthalpy of the plasticised material and varies

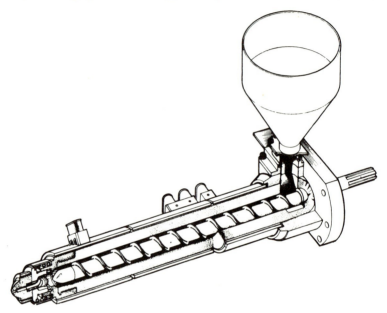

Figure 5.3 Pre-plasticising injection-moulding machine; one cylinder, single screw (courtesy ICI Ltd)

inversely with it, i.e. high-enthalpy material will plasticise less rapidly than low-enthalpy material. Nylon is in the former category and low capacities compared with other materials are usually quoted. Injection-moulding machines are designed so that the time elements shown in *Figure 5.2* are capable of being left to a minimum. Some steps (for instance, injection pressure time and setting time in the time cycle) may be determined by the characteristics of the material being moulded or the shape of the moulding.

Other design features are the injection and mould-locking pressure. Injection pressures must be high for nylon since there is a large pressure drop in the mould and cylinder, and values as high as 1400 kgf/cm$^2$ (20 000 lbf/in$^2$) are often used, particularly in plunger machines as shown in *Figure 5.1*. Injection pressure control is very important; for the nylons, two-stage control is advised where a high

## Conversion processes

injection pressure is followed by a lower holding pressure, both steps being controlled. Typically the holding pressure is about 200 kgf/cm² (3000 lbf/in²) less than the injection pressure.

Moulding machines for nylon require heavy-duty heaters to deal with the high enthalpy, and precise temperature control is necessary. The cylinder should have at least two and preferably three heating zones independently controlled, and the nozzle and mould must have their own control. The latter should also have facilities for cooling, to optimise cycle time. Individual nylon types and grades require fairly precise temperatures in the various areas of the moulding machine, and powder suppliers usually detail these temperatures and ranges for optimum working.

### 5.2.5 MOULDS AND THEIR DESIGN

Mention has already been made that the moulds for use with nylon must have good mating surfaces to avoid flash. Runners and gates should be large to avoid premature solidification and incomplete filling of the mould cavity, and the moulds must incorporate good temperature control. In the mould itself sharp changes of section, which can give rise to stress concentrations in the moulding, must be avoided and radiused bends rather than sharp corners should be used. To determine whether the shrinkage allowance during moulding has been adequate a prototype mould is often used.

Sprues should be as short as possible with a minimum taper of about 2° to 6° included angle. The internal surface should be highly polished and the sprue outlet should blend smoothly with the runner.

Runners should be round and as short as possible to avoid pressure loss during flow, but trapezoidal runners cut into one half of the mould to facilitate ejection are sometimes used. All internal surfaces, in common with the rest of the mould, should be well polished. In multicavity moulding a balanced runner system must be used; that is, the design must be such that as far as possible the plasticised material travels an equal distance from the sprue to the gates.

There are very many types of gate suitable for moulding nylon, but selection is usually determined by the shape of the mould and thickness of mould section. *Figure 5.4*[4] shows a selection of gates commonly used for nylons, while *Table 5.2* lists the characteristics and advantages of those commonly used. Certain general precautions should be observed when designing gates. For example,

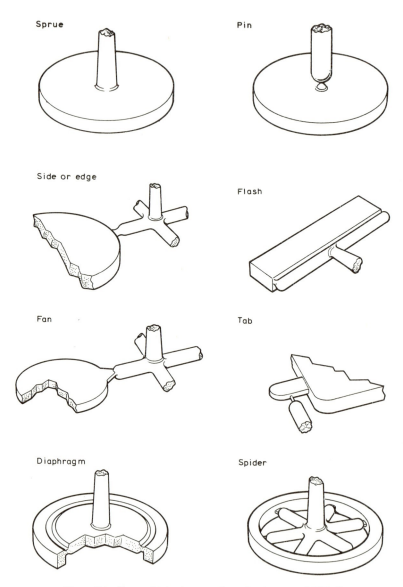

Figure 5.4  Types of injection gate for nylon (courtesy ICI Ltd)

## Conversion processes

gates should be located in the thickest section of the moulding to prevent voids and sinkage. They should be placed away from the area where the highest stress concentration is expected to occur in the moulding in service. Thick sections need large round or rectangular gates, thin sections restricted or fan gates.

**Table 5.2** MOULD GATES FOR NYLON

| Gate type | Characteristic | Suitability |
|---|---|---|
| Sprue | Simplicity, provides symmetry | For single cavity moulds and thick sections |
| Pin | Small cross-section, restricts flow | Automatic degating; minimises finishing; used for thin sections |
| Side (edge) | Simplicity | Multicavity moulds; medium or thick sections |
| Flash | Simultaneous filling over the whole length of the mould to give even shrinkage along the length | |
| Fan | Spreads flow over a large area | Thin section mouldings |
| Tab | Restricted gate, reduces 'jetting' | For mouldings of lowest strain |
| Diaphragm | Like sprue gate | Single-cavity concentric mouldings |
| Submarine (tunnel) | Gate is at an angle to runner | For automatic degating |

Reference has already been made to mould venting, which is important in moulding nylon. The mould should be designed to eliminate or reduce as far as possible blind cavities and convergence of frontal flows in cavities. If these occur, venting should be provided for at the weld lines or at the end of the blind cavities by boring small diameter holes. It may also be possible to vent along injection or core pins (because of the low fluidity and fast filling rates) and, by careful attention to design, natural venting at these points may be adequate.

### 5.2.6 MOULDING DEFECTS AND REMEDIES

Due to the large number of controls required for the variables in the modern injection-moulding process, and the interaction between these variables, the production of mouldings with one or other of a number of defects is on occasion inevitable. Most suppliers of injection-moulding materials include information on moulding faults and remedies in their data brochures and it is instructive to examine these. A typical list is shown in Table 5.3[5].

**Table 5.3** MOULDING DEFECTS; PROBABLE CAUSES AND REMEDIES
(courtesy Polypenco Ltd)

| Moulding fault | Probable cause | Possible remedies |
|---|---|---|
| Short mouldings | 1. Insufficient material injected | 1. Check hopper feed |
| | | 2. Increase feed setting |
| | 2. Poor melt flow | 1. Increase injection pressure |
| | | 2. Increase injection rate |
| | | 3. Increase cylinder temperature |
| | | 4. Increase mould temperature |
| | | 5. Increase vent size |
| | | 6. Increase nozzle/sprue/ runner/gate size |
| | | 7. Check mould design for restrictions |
| | 3. Low material bulk density | 1. Increase feed setting |
| | | 2. Decrease regrind proportion |
| Excessive flash | 1. Insufficient mould locking force | 1. Increase locking pressure |
| | | 2. Decrease injection/holding pressure |
| | | 3. Decrease injection rate |
| | | 4. Decrease melt temperature |
| | 2. Overpacked mould | 1. Decrease feed setting |
| | | 2. Check venting size and efficiency |
| | 3. Mould misalignment | 1. Reclamp mould |
| | | 2. Check mould alignment pins |
| | | 3. Regrind/realign mould mating surface |
| | 4. Mould deflection | 1. Provide backing plates for mould |
| Drooling | 1. High melt temperature | 1. Decrease nozzle temperature |
| | | 2. Decrease cylinder temperature |
| | | 3. Increase cycle time |
| | 2. Excess back pressure | 1. Decrease back pressure |
| | 3. Damp material | 1. Ensure resin is dry |
| | 4. Incorrect nozzle design | 1. Check nozzle temperature variations |
| | | 2. Use reverse taper nozzle |
| | | 3. Incorporate shut-off valve |

**Table 5.3** (*continued*)

| Moulding fault | Probable cause | Possible remedies |
|---|---|---|
| Splash marks, streaks or mica marks | 1. Moisture in granules<br>2. Degraded material | 1. Ensure resin is dry<br>1. Decrease melt temperature<br>2. Eliminate dead spots in cylinder<br>3. Increase material throughput |
| Burn marks or gassing | 1. Inadequate venting<br><br>2. Restricted melt flow<br><br><br><br><br>3. Damp material | 1. Decrease injection rate<br>2. Increase vent size<br>1. Decrease injection rate<br>2. Decrease melt temperature<br>3. Increase gate size<br>4. Increase nozzle size<br>5. Relocate gating<br>1. Ensure resin is dry |
| Voiding or sink marks | 1. Inadequate supply of material to cavity before freeze-off<br><br><br><br><br><br><br><br>2. Insufficient plasticising capacity<br>3. Poor mould design | 1. Increase injection/holding pressure<br>2. Increase holding pressure time<br>3. Increase feed setting<br>4. Increase nozzle temperature<br>5. Increase mould temperature<br>6. Decrease melt temperature<br>7. Increase gate size<br>8. Increase runner/sprue/nozzle size<br>1. Preheat granules<br>2. Change to larger machine<br>1. Regate at thickest section |
| Windows | 1. Unmelted granules visible on mouldings | 1. Increase cylinder temperature<br>2. Increase cycle time<br>3. Increase back pressure<br>4. Preheat granules<br>5. Change to larger machine |
| Brittleness | 1. Damp material<br>2. Melt degraded<br>3. Incomplete mixing of melt<br>4. Inadequate feed<br><br>5. Poor component design | 1. Ensure granules are dry<br>1. Lower cylinder temperature<br>1. Raise cylinder temperature<br>1. Increase holding pressure<br>2. Increase injection/holding pressure<br>1. Eliminate sharp corners or abrupt changes in section |
| Weld lines | 1. Incomplete mixing of melt in cavity | 1. Increase injection pressure<br>2. Increase injection rate<br>3. Increase cylinder temperature<br>4. Increase mould temperature<br>6. Enlarge gate and runners<br>7. Enlarge venting<br>8. Relocate gate |

**Table 5.3** (*continued*)

| Moulding fault | Probable cause | Possible remedies |
|---|---|---|
| Jetting | 1. Melt enters cavity too rapidly | 1. Decrease injection speed<br>2. Change gate design or position |
| Dull surface finish | 1. Mould defects<br>2. Incorrect melt temperature (unreinforced grades)<br>3. Damp material | 1. Polish mould cavity<br>1. Decrease mould temperature<br>2. Increase cylinder temperature<br>1. Ensure granules are dry |
| Poor release | 1. Excessive mould temperature<br>2. Overpacking in mould<br><br>3. Premature release<br>4. Poor mould design | 1. Eliminate mould hot spots<br>2. Lower mould temperature<br>1. Decrease injection/holding pressure<br>2. Decrease hold time<br>1. Increase cooling time<br>1. Check mould taper, undercuts etc.<br>2. Polish mould |
| Warping | 1. Part warps on ejection<br><br><br><br><br>2. Part warps during post-treatment or in use<br><br><br><br>3. Unsymmetrical mouldings<br><br>4. Differential shrinkage of glass-reinforced mouldings | 1. Increase cooling time<br>2. Lower mould temperature<br>3. Decrease holding pressure<br>4. Increase core temperature<br>5. Increase size or number of ejection pins<br>1. Increase mould temperature<br>2. Eliminate temperature differences between mould halves<br>3. Increase cylinder temperature<br>4. Increase cooling time<br>1. Use mould temperature differential<br>2. Reduce changes in moulding section<br>1. Increase cooling time<br>2. Decrease injection pressure<br>3. Decrease holding pressure<br>4. Use mould temperature differential<br>5. Balance mould cavity filling |
| Screw stalling | 1. Melt viscosity too high | 1. Preheat granules<br>2. Increase rear zone temperature<br>3. Increase back pressure |

## 5.2.7 POST-MOULDING TREATMENT

Since dry resin is used for moulding, freshly moulded components are quite dry and, particularly those from nylon 66 and nylon 6, may not necessarily have the optimum mechanical properties or dimensional stability required for their anticipated service. It should be pointed out, however, that the great majority of injection-moulded nylon components are not critical in this respect.

Deficiencies arise from two main causes, the presence of molecular orientation and frozen-in stresses due to unidirectional flow during moulding and differential cooling of the moulding, and the absorption of moisture and consequent dimensional change of the as-moulded component tending to reach an equilibrium moisture state in the service environment. Correct design of components should allow for these effects, and the deficiencies may be alleviated by using annealing and/or moisturising operations after moulding.

### 5.2.7.1 Annealing

This process is aimed at removing moulding stresses and to some extent molecular orientation caused by directional effects due to flow during moulding. The components are immersed in a non-oxidising oil or wax and held within a certain temperature range, according to the nylon type, for a period depending on the maximum thickness of section. For nylon 66 the range is about 160–190°C and the soak time about 15 minutes per 3 mm of section. Slow heating to and cooling from the soak temperature is required, and it is preferable for the moulding to be submerged in the oil throughout the whole cycle. When the maximum service temperature is known to be below the range mentioned above it is usually sufficient to soak the moulding at about 20°C in excess of this known temperature, thus reducing cycle time and temperature. As the annealing process tends to dry out mouldings, those made from nylons that have a higher equilibrium moisture content in air are usually moisturised after annealing.

### 5.2.7.2 Moisturising

This process is usually aimed at increasing the moisture content of the moulding to a value nearer the equilibrium value with respect to the anticipated environment, thus increasing dimensional stability. At the same time increase in moisture reduces brittleness in the moulding, but may also to a certain degree reduce stiffness.

## 5.3 Extrusion

Extrusion may be distinguished from injection moulding by the fact that in the former process the product is of uniform cross-section while in the latter the product, the moulding, can be and usually is of variable section according to the mould design. Extrusion is therefore used for the continuous production of long lengths of shapes in simple sections such as rod, tube, bar, sheet, film etc., while injection moulding is used for the batch production of large numbers of items of unique form where there is little need to perform second operations before the item is ready for use.

### 5.3.1 THE EXTRUDER

Historically, machines of the ram and cylinder type were first used for the extrusion of plastics, and this type has the advantage of better pressure control; however, the discontinuous nature of the ram process led to the development of the screw machine, which is widely used today for the extrusion of polyamides and thermoplastics in general. Fundamentally the screw-extrusion machine consists of a screw rotating in a heated cylinder with a feed opening set at an

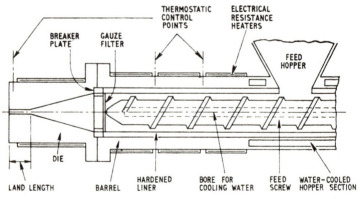

Figure 5.5    Essential features of a screw extruder

# Conversion processes

angle to the axis at one end, and an orifice or die at the other. The die is usually preceded by a breaker plate supporting a screen, which serves to provide back pressure in the cylinder and increase homogeneity in the moving thermoplastic mass. Figure 5.5[6] illustrates diagrammatically the features of a typical screw extruder.

## 5.3.2 EXTRUDER SCREW DESIGN

An extruder screw should be designed to handle only one type of thermoplastic, since the variables are so complex that it is rarely possible with one design to obtain optimum extrusion conditions for several material types. Consideration will be given here to screws found suitable for polyamides.

Polyamides differ from the majority of thermoplastics in having sharp melting points. This implies a sudden solid–fluid transition, so the transition or compression zone in the extruder is usually quite short, sometimes as low as half a flight-turn for nylon 66 or 6.10, but longer for nylon 6.

The design characteristics are usually expressed in terms of the screw length, the cylinder (barrel) diameter, the screw length to cylinder diameter ($L/D$) ratio, the lengths of the feed, compression and metering zones, the compression ratio, and the screw pitch and helix. The channel depths in the feed and metering zones are usually

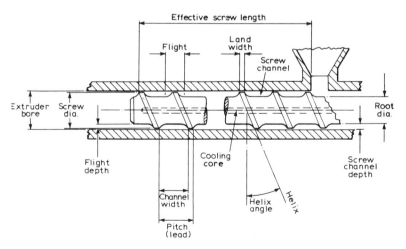

Figure 5.6  Nomenclature of typical single-start extrusion screw (courtesy ICI Ltd)

specified. The definitions of these and many other extrusion terms are given in BS1755. *Figure 5.6*[7] illustrates the application of some of the terms to a typical nylon extrusion screw. Typical values for screw design variables for a single-screw nylon extruder are given in *Table 5.4*. These are more applicable to nylon 66, nylon 6 and nylon 6.10, the design requirements for nylon 11 and nylon 12 being less critical.

Table 5.4 TYPICAL VALUES FOR NYLON EXTRUSION SCREW VARIABLES

| | |
|---|---|
| Length/diameter ($L/D$) ratio | Not less than 20:1 |
| Length of feed zone | $13.5D$ |
| Length of compression zone | $0.5D$ |
| Length of metering zone | Not less than $6D$ |
| Compression ratio | 2.5:1 to 4:1 |
| Screw pitch | $1-1.8D$ |
| Helix angle | $17°-30°$ |
| Channel depth, feed zone | $0.15-0.2D$ |
| Channel depth, metering zone | $0.08D$ maximum |

### 5.3.3 PROCESS VARIABLES

With a fixed screw design, output and quality of extruded material can be controlled by altering melt temperature and screw speed. While these can be independently controlled, the range of suitable working conditions is not particularly wide since for steady output it is essential to have a system in stable equilibrium. For example, too high a melt temperature combined with a low back pressure can lead to surging, or too high a screw speed leads to high shear rates in the melt and melt fracture.

Provision is sometimes made for cylinder cooling in certain zones, but screw cooling as used for some other thermoplastics is not recommended for nylons.

The theory of the screw extrusion machine has been widely developed, and detailed information is given in textbooks by Fisher[6] and Schenkel[8].

### 5.3.4 VARIATIONS IN BASIC SCREW DESIGN

Over the past two decades variations in the basic screw extruder design have been developed to allow less critical working and higher outputs. A few of these are discussed below.

### 5.3.4.1 Vacuum extraction

To facilitate extraction of moisture and other volatiles in nylon after it has been delivered to the screw from the feed hopper, a vented barrel is sometimes used. This is illustrated in *Figure 5.7*[7]. The inclusion of a decompression zone is necessary to allow for the escape of the volatiles, but care must be taken to ensure a correct balance between the metering and compression zones of the screw, in order to avoid blocking the vent. This occurs when the pumping capacity of the compression zone exceeds that of the metering zone.

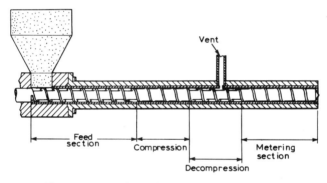

Figure 5.7  Vented barrel extruder (courtesy ICI Ltd)

A further improvement on the vented-barrel extruder is the provision of a by-pass channel between the decompression and compression zones. A valve in the by-pass can be set to control the output from the first screw stage.

### 5.3.4.2 Fast-running screws

These screws have been designed for throughput two to three times faster than is possible with conventional screws. Shaft speeds are between 250 and 2500 rev/min, with speeds at the screw surface of up to 6 m/s (20 ft/s). $L/D$ ratios of 12 or less are used, with a metering zone approximately 2D in length. Only a small amount of material is in the barrel at any one time and the feed zone is never full.

These fast screws permit adiabatic extrusion and, after the initial heating of the feed material by external sources, sufficient frictional heat is developed by the shearing action of the screws to maintain the correct temperature in the melt without the need for external heat input.

### 5.3.4.3 Twin screws

Twin-screw extruders (where parallel, completely intermeshing screws are housed in the same barrel) were developed from the single-screw machines in order to obtain higher pressures and more uniform working. At the same time additives can be incorporated in the mix more quickly and uniformly.

In the twin-screw machine the melt is not so sensitive to back pressure as in the single-screw machine, especially if the clearance between screw and barrel is small. The feed and metering zones can be made smaller and the screw length is typically $7D$ to $12D$. Usually only two control heating zones for the screw are required.

For a given output a twin-screw extruder costs more than a single-screw one, but in certain circumstances the processing advantages outweigh this.

## 5.3.5 GENERAL PROCESSING CONDITIONS FOR NYLON EXTRUSION

### 5.3.5.1 Feed

The raw material for extrusion (the resin) is generally in the form of short cylinders or cubes, but sometimes powder, and the feed opening or throat situated at the base of the conical or rectangular feed hopper is specially designed to take the form of feed used. Care must be taken to feed slowly and to ensure that the hopper is never empty during extrusion. This is particularly important with fast-running or twin-screw operation. Temperature must be low enough to avoid undue softening of the feed near the hopper.

### 5.3.5.2 Pre-heating of feed

Output is increased considerably by preheating the feed material, and at the same time moisture content is reduced. Typically, heating to 80°C can increase output by up to 30 per cent compared to feeding at ambient temperature. If heating is used, it is important to control temperature to ensure constant rate of output.

### 5.3.5.3 Barrel heating

As already mentioned, with single-screw extruders the three main zones should have independently controllable heaters. The heat input required will depend on screw design and the melt viscosity of the nylon type and grade. *Table 5.5*[8] shows typical extruder and die temperatures in various locations of a single-screw extruder for different types of nylon.

**Table 5.5** TYPICAL TEMPERATURES (°C) IN DIFFERENT EXTRUDER ZONES FOR VARIOUS NYLONS (courtesy Carl Hanser Verlag)

*Screw parameters:* $D = 60$ mm, $L/D = 15$–$20$, channel depth $\approx 3$ mm, screw speed = 60 rev/min

| Location | Nylon Type | | |
| --- | --- | --- | --- |
| | 66 | 6, 6.10 | 11, 12 |
| Rear cylinder | 250 | 220 | 200 |
| Middle cylinder | 275 | 250 | 230 |
| Front cylinder | 285 | 270 | 250 |
| Head | 280 | 270 | 240 |
| Die | 275 | 260 | 230 |
| Stock temperature | 275 | 255 | 220 |

### 5.3.5.4 Power requirements

The power input of a screw extruder depends on many factors such as screw diameter, length, detailed design and running speed, and also on feed rate and melt viscosity of the resin. It is usual to relate power input to extrusion output in units of kilowatts per hour per kilogram (or pound). A value of $0.3 \, \text{kW} \, \text{h}^{-1} \, \text{kg}^{-1}$ may be taken as the lowest feasible.

The power increases substantially with screw diameter, and it is common practice in designing to establish equations or plot graphs relating these two parameters.

#### 5.3.5.5 Homogeneity of melt

Homogeneity of the nylon melt is achieved by proper design of the strainer plate, the component that also produces the necessary back pressure in the moving melt. Unmelted or partially melted particles must be removed before the melt reaches the die. One way of increasing back pressure and improving homogeneity is to use a screw that is designed to move axially under controlled conditions. The effect of this is similar to that obtained with a variable-slit device.

#### 5.3.5.6 Melt cooling

Depending on the extrusion form, fluid coolants such as water, organic liquids or air may be used. For some forms cooled metal surfaces, rollers, tubes, etc. are preferable.

#### 5.3.5.7 Change of extrusion material

The procedure adopted when a change of material or change of grade is required is to run the screw, without feed, until the extruder is empty. The cylinder temperature is then reduced by dropping the heat input or by cooling, and the new material is fed in at the lower temperature, and conditions to suit the new material gradually established. In this way possible degradation of the new material is avoided. The change-over period may be reduced by utilising a short term intermediate feed of a low melting material such as a polyolefine or polystyrene. This obviates running with an empty cylinder. Great care must be taken to avoid cooling molten nylon in the extruder screw below its solidification point. Stopping the extruder even for a few minutes can also lead to sudden cooling.

### 5.3.6 TECHNIQUES FOR VARIOUS EXTRUDED FORMS

While details of the extrusion screw and the conversion of the feed have been dealt with in detail in the previous sections, the processes occurring downstream of the screw nozzle determine the shape and form of the finished article. Very many variations of form are possible and the main features of the process for the most common types are described in the following sections.

### 5.3.6.1 Rod and plate extrusion

At first sight the extrusion of good-quality nylon rod would seem to present no difficulties. In the early development of the technique, however, it was found that there was a tendency for the creation of voids in the middle of the extrusion, particularly with thick sections. These voids were created by the shrinkage of the fluid core of the extrusion on solidifying within an already solidified and rigid annulus. The nylons with sharp melting points and high contraction on solidification are particularly prone to this effect. With most other thermoplastics this kind of defect can be overcome by slow and controlled cooling of the extrudate. For nylons the difficulty was overcome by Stott[9], who used a cooled forming-tube connected directly to the die so that the molten core of the extrudate could be pressurised to force the solidified shell against the die wall, forming a partial seal. Since the molten extrudate is continuously under pressure, voids are eliminated. The back pressure created by the expansion of the shell against the die wall can be controlled by the rate of cooling, and equilibrium conditions to give steady extrusion rates can be established. The system is shown diagrammatically in *Figure 5.8*[6].

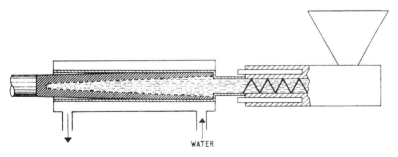

Figure 5.8 System for producing void-free nylon rod extrusions

Using the same principle of forcing a thin solidified annulus of the extrudate against the die wall by the melt pressure, a similar method was patented by NV Onderzoekinginstitut Research[10]. In this method particular attention was paid to avoiding a too sudden cooling of the extrudate on entry to the forming die.

While both the methods described above can produce nearly void-free material, considerable internal stress may still be present in the extrusion, due principally to the differential rate of cooling across the section. For sections over 25 mm (1 in) in diameter it is usual to anneal the material by heating in oil followed by slow cooling. Both the methods described for producing void-free rod are used commercially, and rod up to 200 mm (8 in) in diameter has been produced.

### 5.3.6.2 Tube extrusion

The essential features of a tube extrusion system are the provision of a die with a nozzle of annular shape, and an arrangement for accurately sizing the outside and inside diameter and wall thickness of the extrudate. There are also two basic designs of extrusion head: the straight-through or axial design, where the extrudate proceeds in the direction of the screw axis through the forming, sizing and calibration operations; and the design that uses a crosshead to divert the flow of materials from the nozzle through an angle, usually 90°, past the core pin, set at this angle. In both designs the cross-section of the extrudate assumes the annular form, after which it is sized and calibrated before haul-off. The two designs are shown diagrammatically in *Figure 5.9*[6].

While the straight-through design is generally better for extruding tube of constant wall thickness, it has the disadvantage that the temperature of the die core, or torpedo, cannot be easily controlled due to the difficulty of passing the heat supply through the spider arms. The crosshead die provides easy access to the mandrel and obviates the need for a spider. Heating and cooling of the core is facilitated. Also, in the straight-through die, interruption of the melt flow by the bore support causes weaknesses, which are formed in a continuous line (weld line) behind each support in the extrudate. These may not be visible but exist structurally in the material.

In both designs care must be taken to avoid sudden change in cross-section. As a rule this is gradually reduced on the downstream side of the spider.

The cross-section of the extrudate is seldom the same as that of the die orifice, and the final shape and size are usually determined by a

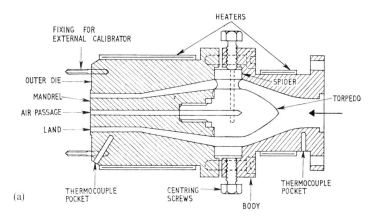

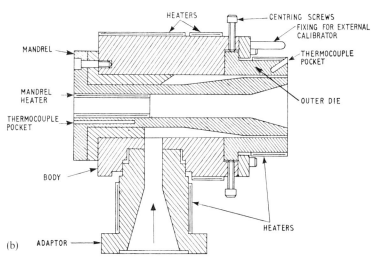

Figure 5.9 Tube-extrusion dies: (a) straight-through; (b) crosshead (courtesy ICI Ltd)

combination of sizing dies and calibration tubes using vacuum (preferably) or air pressure to force the extrudate to assume the contour of the wall of the tube. Pressure is normally used in sizing small-bore tubing and is usually injected through the torpedo to the internal bore of the extruded tubing, preventing collapse and determining size and wall thickness. Vacuum, when used, is applied externally to the tubing as it leaves the die and enters the calibration tube. In the latter case there is usually a gap between die and vacuum system, at which point the extrudate is drawn down slightly.

As the tubing takes up its final size in the calibrator, heat must be removed to prevent further dimensional changes; for this reason the calibrator is usually water-cooled.

High-viscosity grades of nylon must be used for close-tolerance tubing and in all cases where there is an air-gap between extrusion die and calibration unit.

### 5.3.6.3 Wire coating/cable-sheathing

Nylon is not often used as a primary wire coating. Most frequently it serves as an outer abrasion- and chemical-resistant jacket for wire already insulated with some other material, for example PVC.

Low melt-viscosity grade nylons are used for wire coating, and extrusion temperatures are kept fairly high. High drawdown ratios and extrusion speeds are thus possible, and the thickness of the sheath in the cable-sheathing process is controlled by the latter.

A crosshead die such as illustrated in *Figure 5.10*[7] is used for cable covering. To accommodate the variable thickness of the primary insulated cable, there must be sufficient clearance between the core and the torpedo tip.

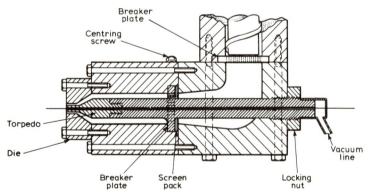

Figure 5.10  Vacuum-assisted cable-sheathing crosshead die (courtesy ICI Ltd)

The whole extrusion process for wire or sheath covering with nylon involves, in addition to the extruder, a sequential arrangement

## Conversion processes

for cooling, haul-off by capstan, and wind-up similar to that used in polyethylene or PVC wire insulation. A linear measuring unit and equipment for testing diameter of the covered product are often included in the line.

### 5.3.6.4 Film extrusion

Nylon film may be extruded as blown film (tubular film) or flat film. For blown film a nylon of high viscosity (about ten to twenty times higher than used for wire and cable covering) is required, and die temperatures are commonly about 5°C higher than the melting point of the nylon.

For blown nylon film a conventional process is used that involves inflating the extruded tube at the die by passing air through the torpedo and quenching the melt by blowing air uniformly over the outside of the inflated tube. The air pressure inside the tube and the haul-off speed determine the thickness and width of the film. After the inflated tube is cooled it is usually collapsed and flattened between nip rolls before winding-up.

Flat film is produced by extrusion through a straight slot die followed by quenching on cold rolls or in water. Medium viscosity nylon grades are used, and die temperatures are commonly about 15–20°C higher than the melting point of the extruding nylon. Because of more efficient cooling the crystallinity of flat cast film is usually less than that of blown film made from the same type of nylon.

### 5.3.6.5 Monofilament

Monofilament is generally considered to be small-diameter, circular cross-section extrusion, limited at the lower end to about 0.1 mm (0.004 in) in diameter and at the upper end to about 1.8 mm (0.070 in) in diameter.

The equipment for the production of nylon monofilament consists basically of an extruder, a drawing/conditioning unit and a bobbin wind-up mechanism. The processing principle is very similar to that of the melt spinning of nylon fibre. The monofilament extruder consists of a gear pump (which replaces the screw of conventional extrusion) and a sand filter-pack placed before the multiple-orifice die. The filament is first extruded into a water quench tank at 40°C placed just below the die. Thereafter the filament passes to two sets of gripping rolls (called godets) running in series at different relative speeds; this, aided by air heating, results in drawing, reduction in diameter and increase in strength of the filament. A tendency for the filament to contract after this treatment is overcome by passing the line through a heated conditioning chamber before final wind-up.

Low-viscosity grades of all the commercial types of nylon are used for production of monofilament.

## 5.4 Melt spinning

For completeness in this chapter, melt spinning must be discussed briefly since it is the oldest and most traditional of the conversion processes for nylon, but it is beyond the scope of this book to cover

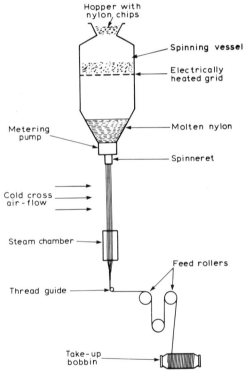

Figure 5.11 Arrangement for melt-spinning process (courtesy ICI Ltd)

*Conversion processes* 173

the very wide subject of nylon textile technology, which is dealt with adequately in other standard works[11,12].

The most common nylons used in melt spinning of fibres are nylon 66 and 6. It is sometimes convenient to manufacture the nylon chips or flake used in melt spinning at a location remote from that where spinning is carried out, as the latter process is best located in a dust-free and clean environment. The nylon polymer for melt spinning must in any case be of the highest purity. Particulate contamination especially can show up in yarn discoloration, and indeed may form points of weakness in the yarn. Compared with nylons intended for plastics uses, polymers used for spinning are of relatively low molecular weight and, therefore, low viscosity.

A typical melt spinning plant is shown diagrammatically in *Figure 5.11*[4]. The sequence of the processing steps is self-explanatory. It must be pointed out that the present-day scale of operations for melt spinning of nylon fibre is vastly greater, about 100 times, than that of melt extrusion.

## 5.5 Casting

Nylon castings may be produced by melting chip or flake nylon in an extruder, passing it into a stationary mould and cooling it under pressure. This method, however, is seldom used commercially and for nearly all casting the method of activated anionic polymerisation from the monomer is used. The chemistry of the process has already been described in Chapter 3.

At the present time the lactams are the only intermediates subject to anionic polymerisation with a nylon as the final product. Only two lactams, caprolactam and laurinlactam, have been used commercially to produce monomer cast nylon.

### 5.5.1 DIRECT CONVERSION PROCESSING

Casting by the anionic polymerisation method is not a conversion process like the others considered in this chapter, where already polymerised nylon in chip, flake or powder form is converted to the semi-finished or final shape required by the end-user. This type of nylon casting is a unique direct conversion from the intermediate to the final form required by the end-user. The form may be stock shapes such as rod, tube, plate, etc., or custom castings where moulds are made up to the customer's requirement. Foundry techniques as used for metals are commonly adopted in cast nylon production—for example, pattern making, mould design and manufacture, pouring or casting, controlled cooling, annealing, to name a few.

### 5.5.2 REQUIREMENTS OF INTERMEDIATES

Activated anionic polymerisation requires as starting material (in addition to the lactam) an initiator, sometimes termed activator, and a catalyst, commonly the sodium salt of the lactam. These two species are in small proportion in the mix. The chemical structure and concentration of the initiator determine the structure, degree of crystallinity, and amount of cross-linking in the polymer. These compound species have been the subject of numerous patents of which the claims generally concern improvement to be gained in the properties of the resulting polymer by use of the initiator, or increased chemical stability in the system[13].

It is essential that the starting materials in anionic polymerisation are of the highest purity. Water and acidity must be rigorously excluded from the lactam. A distillation step is sometimes used to remove the last traces of moisture from the lactam in process, and is usually carried out under vacuum. Since the melting point of caprolactam is only 69°C it is processed in the liquid phase. A blanket of dry nitrogen is usually used to prevent moisture ingress during processing.

Below 150°C activated anionic polymerisation is slow and no useful polymers are formed, but above this temperature reaction proceeds exothermically and temperature control is essential to obtain good-quality strain-free castings. Some initiators have a fairly long induction period before reacting; others with a short induction period can cause trouble, particularly when filling large moulds, because incomplete mixing with the monomer may occur resulting in non-uniform polymerisation and property variations in the casting.

### 5.5.3 COMPONENT MIXING

The components of the mix are generally held in separate vessels in readiness at the correct temperature until just before casting, but, to reduce the problem of intimately mixing three components, separate two-component mixes may be prepared some time beforehand. Reaction is negligible unless all three components are present. One of two systems may be used to obtain the three-component reactive mix. Two separate, but approximately equal, reservoirs of the molten lactam are held in readiness; one reservoir contains the initiator, the other the catalyst in the correct quantity and at the correct temperature for optimum reaction. At the time of casting, approximately equal volumes of the two component mixes are delivered through separate metering valves to a mixing head and pumped into the moulds. It is claimed with this system that mixing is faster and more intimate and that greater throughputs may be achieved. When it is considered that once mixing has commenced

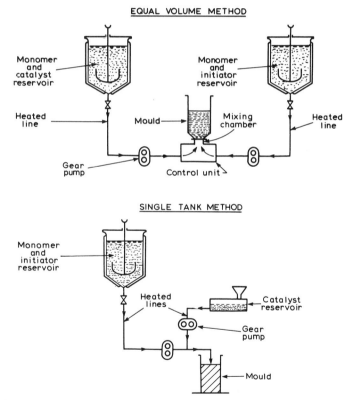

Figure 5.12 Activated anionic polymerisation of lactams: component mixing methods

the polymerisation may be complete inside a few minutes, it is clear that fast mixing and pumping to the mould are essential.

In the alternative system of mixing, a reservoir of lactam, containing the correct quantity of initiator for optimum reaction and at the correct reaction temperature, is pumped directly to the mould, and catalyst from a separate vessel is injected into the main stream through a metering valve. This system has the advantage of flexibility since it may be necessary to alter the concentration of the catalyst during the period of mould filling. The two systems are illustrated diagrammatically in *Figure 5.12*.

### 5.5.4 ABOVE-THE-MELT POLYMERISATION

While nylon casting as practised commercially is a batch process in which one or several moulds are filled sequentially from the same reactive mix, several patents have been filed[14] that claim a method of continuous casting. In one case the reaction temperature is taken above that of the melting point of the resultant polymer (in the case of polycaprolactam 210–230°C), reaction being completed within an extrusion screw; the product may then be extruded conventionally to form rod, tube or other profiles. This method is in contrast with batch casting, which is usually carried out below the melting point of the polymer.

### 5.5.5 CASTING TECHNIQUES

The two techniques most widely used are stationary casting, where the moulds are filled in a stationary position, and centrifugal casting, or spinning, where the fluid mix is subjected to centrifugal forces while the mould rotates at high speed on one or two axes. Stationary casting is generally used for stock shapes such as large diameter rod, thick-wall tube and heavy plate, while thin-wall tube is best produced by spinning. Many custom castings of irregular shape are often produced by spinning since the centrifugal force acting on the fluid ensures that all corners and crevices of the mould are filled. The mould design must ensure that air entrained in the polymerising fluid can be displaced to the central axis to allow venting to the fluid surface.

## 5.5.6 MOULD DESIGN

As in metal foundry work, mould design must be carefully examined especially for parts where close tolerances must be held. Allowance must be made for post-solidification moulding shrinkage, which for unfilled polycaprolactam is a linear 3–4 per cent. Sudden changes in section in the moulding raise special problems because, once cast, the rate of heat loss during cooling is greater from the surface of the thinner section. If the section is asymmetrical, differential cooling rates can give rise to distortion in the casting or residual stresses that may not be apparent immediately but are liable to distort the material at some time after it has been put into service.

## 5.5.7 POST-TREATMENT OF MOULDINGS

To eliminate residual stresses in freshly-produced castings slow cooling of the casting immediately after polymerisation is recommended, especially where thick sections are involved. The alternative is oil-annealing, which is the recommended method if the castings have been first allowed to cool to room temperature and still retain residual stresses. Oil-annealing is a relatively expensive and time-consuming method of removing stresses, since the thermal conductivity of nylon is relatively low and the heating and cooling phases of the cycles must be prolonged to avoid over-stressing the material thermally. In the process the temperature of the oil tank containing the material for annealing is raised slowly to above the glass transition temperature of the polymer, maintained at a constant temperature for several hours, then allowed to cool slowly. The oil increases the rate of heat transfer in the process and is suitably selected to reduce surface oxidation of the casting to a minimum.

## 5.5.8 SPECIAL FORMULATIONS

Addition of special ingredients to the reactive lactam mix is readily carried out to produce cast material of modified properties. The additives may react chemically with one of the main components of the mix (generally the initiator) to alter reaction velocity and indirectly the mechanical properties of the final cast. Foam-producing additives are also of this category[15].

Examples of additives that are chemically non-reactive in the lactam mix, but are added to confer specific properties to the cast material, are molybdenum disulphide for wear resistance, glass spheres or fibres for increased stiffness, and non-reactive organic plasticisers for increased impact resistance.

## 5.6 Powder processing

There are a number of special processing techniques that are adopted to use powdered polyamides. Four of these are used commercially: (a) fluidised-bed coating, (b) flame spraying, (c) electrostatic coating, and (d) pressing and sintering.

### 5.6.1 FORMS OF POWDER

In all these techniques the nylon is used in a special powder form in which the type, size and size distribution are closely controlled. Practically all the powders used for coatings are produced from the commercial nylons 66, 6, 6.10, 11 and 12. The two latter, because of their superior flow properties and lower melting point, have found widespread use.

Coating powders are usually obtained by feeding granular powder into some form of impact mill, after cooling to well below 0°C using liquid nitrogen. High throughputs are possible, producing powder of particle sizes less than 300 μm. A typical size distribution for nylon 12[16] is:

| Size (μm) | Percentage |
|---|---|
| < 60 | $15 \pm 10$ |
| 60–100 | $35 \pm 10$ |
| 100–150 | $35 \pm 10$ |
| 150–200 | $15 \pm 10$ |
| > 250 | < 5 |

The powders for fluidised bed coating are usually heat- and weather-resistant grades. Powders for pressing and sintering are obtained by precipitation from hot glycol solutions of the nylons. This yields a product with a high percentage of crystalline (about 80 per cent), which imparts wear resistance to the final sintered product. The particles obtained in the filter cake from the re-precipitation process are dried, coarsely ground and sized. Thus the powder used for sintering is composed of agglomerates of the particles originally obtained in the precipitation. A typical size analysis of such a nylon 66 powder is given in *Table 5.6*[17].

Table 5.6 TYPICAL SIZE ANALYSIS OF NYLON 66 POWDER FOR SINTERING (courtesy *Modern Plastics*)

| NBS screen size | Weight percentage |
|---|---|
| > 12 | 0 |
| 12–20 | 13.9 |
| 20–40 | 29.1 |
| 40–70 | 23.3 |
| 70–140 | 14.1 |
| < 140 | 19.6 |

## 5.6.2 FLUIDISED BED COATING

In the fluidised bed technique air is passed into the resin powder via a porous sintered metal or glass frit partition. At a certain critical flow rate, vortices are formed in the mass of powder and the bulk volume increases by about 30–40 per cent, at which stage the powder takes on the characteristics of a liquid. If now the degreased or sandblasted metallic components to be coated, heated to between 250°C and 450°C according to the nylon type, are immersed in the fluidised bath for a few seconds, removed and cooled, a solid pinhole-free coating of uniform thickness will have formed on the part. The arrangement used for fluidised bed coating is shown diagrammatically in *Figure 5.13*[18].

For a given temperature the thickness of coating increases with the thickness of the part and the immersion time, but as a rule coating is carried out at the lowest temperature that will produce an even coating and good adhesion of the resin to the base metal. Primers are sometimes used to increase adhesion. Lower temperatures are particularly important where the part to be coated has a large section, as otherwise degradation and blistering due to the elevated temperature may result. With thick sections the technique of 'shock' heating may be used whereby the surface of the part is exposed to a very high temperature for a much shorter time than normal. The heat is largely located in the surface layers of the part; cooling by interchange with the adjacent resin, and the simultaneous resin fusion to form the correct thickness of coating, can be determined by trial and error.

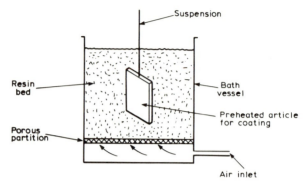

Figure 5.13 Arrangement for nylon fluidised-bed coating process (courtesy Carl Hanser Verlag)

The temperature to which the part is normally heated for fluidised bed coating is within certain ranges according to the resin used. *Table 5.7* shows the ranges for the common nylon types.

**Table 5.7** PREHEAT TEMPERATURE RANGE FOR FLUIDISED BED COATING

| Nylon type | Normal temperature range (°C) |
|---|---|
| 6 | 260–430 |
| 66 | 340–400 |
| 6.10 | 300–400 |
| 11 | 250–380 |
| 12 | 270–330 |

### 5.6.3 FLAME SPRAYING

Many variations of the technique are known. In one variant the powdered resin is carried centrally in an air jet through the nozzle of a spray gun. The jet is heated peripherally at the nozzle by burning propane or oxygen-acetylene mixture delivered through a separate line in the gun, and directed on the metal part to be coated. The temperature of the jet is adjusted to give the correct coating thickness. The equipment for flame spraying can be made portable and

## 5.6.4 ELECTROSTATIC COATING

The method of electrostatic coating developed jointly by the firms of Organico and Sames can yield durable pore-free coatings of nylons, in some cases much thinner than is possible using the fluidised bed or flame-spraying method. In this process the powder is conveyed by means of compressed air to a spray gun where, under the influence of a strong electrostatic field, the material picks up a negative charge and is thereafter directed on to the clean, heated component where it is subsequently fused by further heating.

## 5.6.5 PRESSING AND SINTERING

The processing of nylon powder by pressing and sintering resembles powder metal processing, and indeed sintered nylon products can often replace sintered metal products of similar design. The sintering process is competitive with injection moulding for the production of small parts, and although the latter produces parts having greater tensile and impact strength, and resilience, the former has the advantage that the dimensional stability of the product is invariably superior. Sintered parts subject to wear also tend to be less affected than corresponding injection mouldings.

Patents covering the process have been filed in the US[19]. The process is briefly as follows. The dry powder is compacted in a hardened steel die at a pressure between 24 and 40 kgf/mm$^2$ (15–25 tonf/in$^2$) using a mechanically or hydraulically operated ram. Since the fill ratio is greater than for most powdered metals the die is considerably longer than for these cases. Only a small clearance, about 0.013 mm (0.0005 in), is used between ram and die wall, and it is unnecessary to use draft angles for removal of the green-pressed part. For each formulation an optimum green-density is specified, and pressing is continued until this is achieved. Subsequent sintering of the green-pressing must be carried out under non-oxidising conditions, so a high-temperature resistant heat-transfer oil is commonly used. To obtain optimum results sintering temperatures must be very accurately controlled and sintering cycles, which vary with the section thickness, must be adhered to. A typical oil sintering cycle for sintered nylon 66 powder is 2 hours heating to 257°C, 30 minutes dwell at 257°C, and 2 hours cooling to 90°C.

The die design precludes the direct production of parts with undercuts, and in general tolerances on parts are about the same as those on injection-moulded parts of comparable geometry. Dimensions perpendicular to the direction of ram movement are more easily controlled than those at right angles to this direction.

Sintered nylon is used primarily in bearing and wear applications where its superior frictional properties and higher compressive strength (compared with injection-moulded material) may be used to advantage. A comparison of the critical properties of the two types of materials is given in *Table 5.8*.

**Table 5.8** COMPARISON OF PROPERTIES OF SINTERED AND INJECTION-MOULDED NYLON 66

| Property | Test method ASTM | Sintered 66 | Injection-moulded 66 |
|---|---|---|---|
| Tensile strength $kgf/cm^2$ ($lbf/in^2$) | D648 | 250 (3500) | 790 (11 200) |
| Elongation, per cent | D648 | Nil | > 90 |
| Compressive strength, $kgf/cm^2$ ($lbf/in^2$) | D695 | 1100 (15 700) | 700 (10 000) |
| Flexural strength, $kgf/cm^2$ ($lbf/in^2$) | D790 | 450 (6400) | 970 (13 800) |
| Deformation under load of $140\,kgf/cm^2$ ($2000\,lbf/in^2$) at 50°C | D621 | 0.37 | 1.4 |
| Hardness, Durometer D | D676 | 80 | 78 |
| Specific gravity | D792 | 1.11 | 1.14 |
| Wear in 24 hours (comparative values) | | 1.5 at 1337 (46 800) $PV^*$ | 1.6 at 197 (6900) $PV^*$ (failed in ½ hour) |

\* $PV$ in $N\,m^{-1}\,s^{-1}$ ($lb\,in^{-2}\,ft^{-1}\,min^{-1}$) units.

## 5.7 Machining and finishing operations

All the machining operations used in ordinary metal working (e.g. turning, milling, drilling) can be applied to the nylons, but the tools and their setting, speeds, feeds etc. must be adapted to account for the differences in properties between metals and plastics—in this instance, polyamides. In this context the main properties that cause

## Conversion processes

differences in machining behaviour are thermal conductivity, melting point and stiffness.

The low thermal conductivity of polyamides compared to metals reduces the rate of dissipation of the heat generated between the tool and the workpiece and tends to create local overheating of the machined part.

The melting point of nylons, although considered high for thermoplastics, is very much lower than that of any of the common structural metals. Thus the heat generated in machining tends to degrade, soften and even melt the nylon.

The room-temperature stiffness of unfilled nylons is very much less than that of any of the metals commonly machined. At the elevated temperatures often generated by machining, the stiffness is even less; full support must therefore be given to the workpiece, and low tool pressure employed when machining components where tight tolerances must be held.

Despite the disadvantages mentioned there is no real difficulty in machining any of the nylons to a close tolerance (0.2 per cent of nominal) and with good surface finish. In considering tolerances account must be taken of residual stress in the component due to machining, slow stress relaxation subsequent to machining and, equally, post-machining growth or shrinkage due to moisture absorption or loss. Operations such as annealing and/or moisturising before or after machining, or machining at some partially-finished stage, are frequently carried out to stabilise the machined dimensions of the part. These operations have already been discussed in section 5.2.7.

Practical details for individual machining operation for the nylons are described in the following sections.

### 5.7.1 GENERAL

Tools should be kept sharp and good clearances used. The best results are obtained with very high spindle speeds and low feed rates. Provision must be made to prevent flexing of the workpiece away from the cutting tool. It is desirable to use coolants (preferably of the soluble oil type) to avoid excessive heat generation, even with the harder, higher melting point nylons such as nylon 66 and nylon 6. Coolant also greatly improves the surface finish. Glass-fibre filled nylon types are stiffer and less resilient than unfilled nylons. For the former certain operations such as reaming and filing are easier. On the other hand tool wear is increased due to the presence of abrasive filler.

### 5.7.2 TURNING

Ordinary metalworking lathe tools may be used for the nylons, and for unfilled types a good carbon steel is satisfactory. Automatic screw machines are quite suitable for nylon components, with production rates of up to 1800 per hour.

Turning should be performed by box tools if possible to reduce deflection. The cutter should be set slightly below centre and ground to a radius with a negative rake of 0–5 degrees. Feed rates are generally between 0.1 mm (0.004 in) and 0.25 mm (0.01 in) per spindle revolution, but this should be reduced towards the end of tool travel to obtain a clean surface. Speeds should be in the range 180–300 m/min (600–1000 ft/min).

### 5.7.3 MILLING

Standard milling equipment is satisfactory for nylons and conventional cutters can be used with surface speeds in excess of 300 m/min (1000 ft/min). Climb milling is preferred to reduce deflection of the workpiece during machining, but vertical milling can be used using fly cutter and end mills. The workpiece should be supported along its entire length to reduce deflection.

### 5.7.4 DRILLING

This operation requires the greatest care, as the tendency to heat build-up is greater than for other machining operations. A slow spiral drill with polished flutes is preferred since the large flute area allows free discharge of the chips. The drill should be ground to a point angle of 80–90° with a lip clearance of 9–15°. A general-purpose drill can also be used with an included point angle of 118° and lip clearance of 10–15°. The lip rake should be ground off and the web thinned.

When drilling a series of small holes close together a pin should be inserted in each hole after drilling to preserve rigidity in the workpiece.

Drilling speed to be used should vary inversely with the length/diameter ratio of the hole to be drilled, i.e.

| L/D | Speed |
|---|---|
| 1 | No limit |
| 3 | 45 m/min (150 ft/min) |
| 6 | 30 m/min (100 ft/min) |

A light feed with a 'woodpecker' action of the drill is recommended to facilitate clearance of the swarf and cooling of the bit. This is most important in drilling large-diameter (over 50 mm (2 in)) rod, where excessive heat can lead to fracture and even shattering of the rod. (Special procedures are used for drilling large-diameter rod; precise details are given in manufacturers' data sheets[20].)

It has often been found that oversize drills are required to obtain accurate sizing of holes. This arises from the increased flexibility of the nylon at the temperature of drilling compared to that at room temperature.

## 5.7.5 THREADING AND TAPPING

Since nylon is notch-sensitive like other plastics, a thread with a rounded root should be specified for maximum strength.

Standard taps and dies used for metals will produce clean threads in all nylons, but it is advisable to chamfer the hole preparatory to tapping to avoid damaging the threads if this is carried out as a final stage. High-speed taps from 0.05 mm (0.002 in) to 0.13 mm (0.005 in) oversize should be used as a rule, and coolant is necessary.

## 5.7.6 GRINDING, SAWING AND FILING

The nylons can be ground using a soft-grade wheel with an open grit. Ordinary emery wheels tend to become quickly clogged with ground nylon. A coolant, usually water, sometimes incorporating rust inhibitor, is quite satisfactory and this must be directed continually on the surface to be ground. Nylon rod stock can readily be reduced in diameter by centreless grinding carried out at a speed dependent on the rod diameter.

For sawing nylons band-saws are more satisfactory than circular saws, since the linear nature of the band-saw allows greater heat dissipation. For thicknesses over 50 mm (2 in) it is advisable to cut the nylon on a milling machine or use a powered hacksaw, with copious coolant and adequate blade clearance. Blades with fewer teeth/inch are preferred, and these must be kept sharp to prevent gumming and freezing in the middle of a cut.

### 5.7.7 BLANKING

Blanking of discs, washers and other shapes of nylons is commonly carried out using strip stock up to 3 mm ($\frac{1}{8}$ in) thickness. Both die and punch should be made from steel hardened to Rockwell C62 or C63 and clearances of about 0.013 mm (0.0005 in) should be used. To obtain sharp edges from the thicker strip, hollow knife-edge type dies are preferable. Cracking of stamped parts can be prevented if the strip is preheated by immersion in hot water up to 70°C for 5 to 10 minutes.

### REFERENCES

1. Ogorkiewicz, R. M. (ed.), *Thermoplastics: effects of processing*, Plastics Institute Monograph, Newnes–Butterworths (1969)
2. Walker, J. S. and Martin, E. R., *Injection moulding of Plastics*, 62, Plastics Institute Monograph, Newnes–Butterworths (1966)
3. Ref. 2, 8
4. Imperial Chemical Industries Ltd, *Injection moulding and finishing of Maranyl nylons*, Plastics Division Technical Service Note N 102
5. Polypenco Ltd, *Injection moulding manual*
6. Fisher, E. G., *Extrusion of plastics*, 3rd edn, Plastics and Rubber Institute Monograph, Newnes–Butterworths (1976)
7. Imperial Chemical Industries Ltd, *Extrusion of Maranyl nylons*, Plastics Division, Technical Service Note N 103
8. Schenkel, G., *Kunststoff Extrudiertechnik*, Carl Hanser Verlag, Munich (1963)
9. Stott, L. L., US Patent 2 719 330 (13.3.1951)
10. NV Onderzoekinginstitut Research, British Patent 774 430 (31.5.1953)
11. Moncrieff, R. W., *Man-made fibres*, 6th edn, Newnes–Butterworths (1975)
12. Mark, H. F., Atlas, S. M. and Cernia, E. (eds), *Man-made fibres, science and technology*, Interscience (1968–1969)
13. Monsanto Chemical Co., British Patents 872 328 (5.7.1961) and 916 093 (16.1.1963); Badische Anilin und Soda Fabrik AG, British Patent 900 150 (4.7.1962); Otto Wichterle, British Patent 903 261 (4.7.1962)
14. Otto Wichterle, British Patent 904 229 (22.8.1962); Badische Anilin und Soda Fabrik AG, British Patent 919 246 (20.2.1963)
15. Gilch, H. and Schnieder, K., *Kunststoffe*, No. 1, 13 (1969)
16. Chemische Werke Hüls AG, *Vestamid Polyamide 12*, technical brochure
17. Stott, L. L., *Modern Plastics* (September 1957)

18 Vieweg, R. and Muller, A., *Kunststoff Handbuch*, Vol. VI, *Polyamide*, 362, Carl Hanser Verlag, Munich (1966)
19 Polymer Corporation, US Patents 2 639 278 (19.5.1953), 2 695 425 (30.11.1954), 2 698 966 (11.1.1955), and 2 742 440 (17.4.1956)
20 Polypenco Ltd, *Machining instructions for Polypenco plastics*, Product Data 8, Technical Bulletin

# 6

# Applications of polyamides

## 6.1 General

The general allocation of the polyamides to end-users' applications has been shown in *Table 1.3*. This allocation is to various industrial and commercial areas, and the annual publication of such figures helps in appreciating the trends in polyamide usage in both the UK and the US.

The classification however is inadequate to describe the wide range of applications of polyamides in their many types and forms. Apart from long-established applications there is continuing advance in many fields. An important example is the direct replacement of metals by nylon in components or assemblies of established design, where a cost advantage can be proven with the substitution. New designs of hitherto all-metal assemblies are also being tried out in which the special combination of properties possessed by nylons is used to full advantage and matched to the performance specified.

The improvement in mechanical, physical and chemical properties of nylon, either by modification of the basic molecule (for example by incorporating aromatic segments in the polymer molecules) or by the use of additives to the polymer, is constantly being pursued with a view to serving some new application or extending an existing one.

Striking improvements in mechanical properties can be obtained by incorporating glass-fibre with the basic nylon resin; fibre can be added in high percentage loadings (sometimes exceeding 40 per cent by weight of the composite) to give commercial moulding materials. The mouldings from such materials have a much wider field of application than those from the unfilled resin. By the addition of glass-fibre, the nylon is converted from a tough material to one that is strong and rigid, with mechanical properties resembling those found in structural metals. At the same time creep resistance is increased and heat deflection temperature increased, thereby allowing the high melting point of the basic nylon to be effectively exploited. The fact that this improvement can be achieved without

## Applications of polyamides

sacrifice of electrical, frictional or chemical properties puts glass-fibre filled nylon in a rather special class.

The classification that we shall now adopt for applications of polyamides is based on the main properties utilised in the application, e.g. mechanical, electrical and chemical. It is expedient in the first instance to summarise the properties of nylon that give it an advantage over other thermoplastics. These are tabulated in *Table 6.1*. The application of these properties in end-products is considered later in this chapter.

Table 6.1  PREMIUM PROPERTIES OF NYLONS

| Group | Properties |
|---|---|
| Mechanical | High strength and modulus |
| | Good impact-resistance and toughness |
| | Low friction |
| | Good wear-resistance |
| Physical (including electrical and thermal) | Low specific gravity |
| | High second-order transition |
| | High and sharp melting point |
| | High heat deflection temperature |
| | High specific resistance |
| | Low permittivity and loss factor |
| | Absence of tracking |
| Chemical | Resistance to dilute mineral acids and alkali from pH 4–14 |
| | Excellent resistance to most organic solvents other than a few polar compounds, e.g. phenol, chloroform |

## 6.2 Mechanical applications

### 6.2.1 GENERAL ENGINEERING

In the form of engineering components nylon is making great inroads into the fields traditionally occupied by metals. Parts that previously would have been designed in steel, brass or light alloy can now, in many cases, be made from injection mouldings or from pressed and sintered nylon. In the small component field the techniques mentioned are suitable for rapid production of parts, and to the resulting lowering of costs may be added the benefits usually associated with nylon—that is, strength with lightness, resilience, and freedom from corrosion and noise in service.

## 6.2.2 GEARS AND BEARINGS

Examples of small nylon components now widely used are spur and pinion gear trains, sleeve bearings and bushings, cams and actuators, ball-bearing cages, pistons, and small-bore pipe fittings. *Figure 6.1* shows a variety of such components.

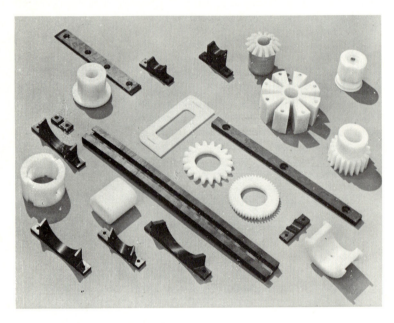

Figure 6.1 Miscellaneous small engineering components in nylon (courtesy Polypenco Ltd)

Nylon gears and bearings are commonly used in small mechanisms where feed lubrication is difficult and a sealed unit may have to be used. When it is undesirable to lubricate at all (for example, in some operations in processing food or textiles) unlubricated nylon gears are a logical choice provided the working temperature is not above about 100°C.

Like their metal counterparts, plastic gears and bearings can be designed using standard engineering formulae and the property data appropriate to the plastic. Due to the low rate of wear of nylon, failure of gears is almost always due to fatigue, and calculations

## Applications of polyamides

have to be made to determine the safe bending stress of teeth. For plain plastic bearings or bushings several National standards exist covering selection of material, component tolerances, and design recommendations[1]. A number of fabricators of engineering plastics

Figure 6.2 Engineering components in sintered nylon (courtesy Polypenco Ltd)

produce their own design manuals[2] for bearings and gears, covering material selection, design of parts and applications. In designing nylon bearings due consideration is given to the need for adequate clearances. Maximum permissible $PV$ ratings, wear rate and wear life at these ratings should have been established.

Small components are generally injection-moulded in moulds designed to produce the final article with a minimum of machining. Components with straight sides can alternatively be produced in sintered nylon. This process can often be automated and used for mass production of parts, e.g. for the automobile industry. The material advantages of sintered nylon have already been discussed in section 5.6. *Figure 6.2* shows a variety of sintered nylon parts.

Large components above a certain size and those of complicated shape are preferably machined from extruded stock shapes or castings.

### 6.2.3 FRICTION AND WEAR APPLICATIONS

The flexibility, wear resistance and low friction of nylon is utilised with advantage in rollers, tyres and covers, and wear plates, and there are few industries where nylon in one form or another is not used to

Figure 6.3  Cast nylon conveyor buckets used in the mining industry (courtesy Polypenco Ltd)

reduce wear or sliding friction. Specific examples are rollers for conveyors in the food industry, drawer rollers and door runners for furniture, tyres for small trolleys and trucks in transport, back-up rolls in the paper industry, and straightening dies in the wire industry. Conveyor buckets in cast nylon are becoming popular. Apart from wear and corrosion resistance, the low friction facilitates almost complete discharge of the bucket contents on tipping. *Figure 6.3* shows a bucket application in the mining industry.

### 6.2.4 ROTATING COMPONENTS

The combination of ease of manufacture, lightness, toughness and corrosion-resistance of the nylons recommends them as useful candidates for rotating members such as fans, propellers and impellers. For air-circulating fans the low-inertia blades can be as efficient as those made in steel or light alloy, with the added advantage of impact resistance. This is useful where misalignment can cause fouling of the blades on static parts of the assembly, as in automobile engine cooling fans.

Nylon is finding increasing favour for small coastal craft propellers, which carry a high risk of impact with floating debris or submerged obstacles. With nylon in these cases a catastrophic failure is less likely than with metal. Increased corrosion-resistance of nylon in sea water is an additional advantage over conventional blade material such as bronze, and it has been shown that cavitation erosion of nylon blades is small. Propeller blades in nylon tend to be designed with thicker sections than their metal counterparts to compensate for the lower stiffness of nylon. High-modulus nylon 66 or monomer cast nylon is the preferred type for blades where stiffness is important. For very high efficiencies where flutter must be kept to a minimum nylon propellers reinforced with fibrous fillers have been chosen to obtain the specified performance.

Nylon impellers for pumps handling abrasive and corrosive fluids have, in many cases, been found to outlast metal impellers subjected to the same service. Here again cast or fibre-reinforced nylons may be used for increased stiffness.

### 6.2.5 HOUSINGS AND CASINGS

A large area of application for the nylons is in component housings for industrial tools, domestic appliances and the like, particularly where a combination of high-temperature resistance, impact resistance and electrical insulation is required and where the overall properties are superior to the thermosets (such as phenol-formaldehyde) traditionally used in these applications.

A wide choice is available. Generally glass-filled nylon 66 is used for housings mainly because of its greater rigidity, improved dimensional stability and higher deflection temperature. Unfilled nylon 66 is preferred where toughness is more important, while nylon 11 or 12 is used where flexibility is required. For large housings monomer cast nylon is the preferred choice.

Small and medium sized housings can usually be injection-moulded while large housings are often centrifugally cast. Examples are electric motor casings for portable power tools (these are often made in clam-shell mouldings), grips and handles for portable tools, housings for static or portable lamps, battery boxes (using alkaline electrolytes), distributor covers for automobiles, vacuum cleaner motor housings, flash gun casings and lens shields.

## 6.3 Electrical applications

### 6.3.1 ELECTRICAL ENGINEERING

Under certain conditions the electrical properties of polyamides are inferior to those of polyolefines and polystyrene, but with superior mechanical properties and heat-resistance the nylons have a particular field of electrical application where this combination of properties is required. Consideration must be given to the environment of the nylon component in service and the possible reduction in useful electrical properties with moisture-uptake. The correct choice of nylon type (e.g. nylon 11 or nylon 12 rather than nylon 66) can offset to some extent the reduction in performance that may arise through moisture absorption.

In addition to their use in electric motor casings and housings as previously mentioned, nylons are widely used as the insulators in electric plugs, sockets and switches, for press-buttons and tumbler components of the latter and for handles of live-line tools. Armatures, insulators, coil formers, clips for cables carrying electric current, or for twin or multiflex electric wiring, are commonly made in nylon, while cable is sometimes sheathed in nylon to resist abrasive wear or chemicals. Nylon rollers are favoured for the components in electromechanical switches because of their resistance to heavy shock and abrasive wear. Sintered nylon also has been used in this application. Finally, nylon screws and washers are widely used for attaching conductors that require insulation from one another.

### 6.3.2 APPLIANCES

The electrical insulating properties of nylon, allied to its toughness, are used widely in components for numerous appliances used in offices, the home, sport, leisure and entertainment. Appliances such as tape recorders, electric food mixers, electric carving knives, hair dryers, washing machines, lawn mowers, radio and television receivers and transmitters use clips, connectors, switch components, spacers and other components (as well as housings, previously mentioned) moulded in nylon.

### 6.3.3 TELECOMMUNICATIONS

The rapid growth in telecommunications since the Second World War has led to the widespread use by large companies and public corporations in the telecommunications field, such as the Post Office, of systems incorporating small components moulded from nylon, e.g. relay coil formers and tag blocks. Glass-filled nylon 66 is preferred because of rigidity. Ease of moulding and resistance to molten solder, properties possessed by nylon 66, are other prerequisites.

## 6.4 Chemical applications

### 6.4.1 TUBING

With due regard to the compatibility of the various types with chemicals, extruded nylon is used widely in the form of tubing, for example as thin-walled tubing for low-pressure fluid lines or in nylon-cored and reinforced hose for high-pressure hydraulic lines.

As a specific example, nylon tubing is used for the air and ink lines on newspaper presses. Tubing with rated burst pressures in the range 40–180 kgf/cm$^2$ (600–2500 lbf/in$^2$) has been used successfully in this application, replacing copper tubing previously used. Superior vibration resistance and flex-fatigue was found using the nylon replacement. Nylon high-pressure tubing has been equally successful as supply lines for pneumatic and hydraulic portable tools and fixtures. In addition to lightness and corrosion resistance it is advantageous that nylon tubing can easily be coloured for identification purposes, and its flexibility to a large extent obviates the need for fixtures and jigs to support the lines. In these applications it is usually possible to make a choice of material between nylon 66 or nylon 6 (essentially stronger, preferred for higher pressures, but less flexible) and nylon 11 or nylon 12 (less strong, but more flexible and in most cases with better chemical compatibility).

## Applications of polyamides

In the automotive industry nylon (mainly nylon 11 and nylon 12) is used widely for petrol lines, control cable sheathing, oil sump overflow lines and distributor-to-carburettor vacuum control lines. *Figure 6.4* shows a typical application of bonded nylon hose in an automobile.

Nylon tubing has also found applications in the food industries for syrup and beverage lines, and in the chemical process industries

Figure 6.4  Automobile application of reinforced nylon hose (courtesy Polypenco Ltd)

for caustic lines. Mention has already been made of the possible embrittlement effect of zinc salts in contact with nylon 66, and prospective users must seek expert advice regarding their particular application.

Where high line-pressures are used in conveying and pumping fluids used in air-conditioning, reinforced nylon hose has been found successful. One design comprises a nylon inner core reinforced with braided nylon yarn and an abrasive-resistant polyurethane external

## Applications of polyamides

cover. The nylon braid is chemically bonded to the seamless nylon core and the external cover bonded to the braid. This type of hose has replaced conventional rubber hose with advantages in weight, bulk and cost. A further advantage is the lower diffusion rate of organic refrigerants through nylon compared to rubber (one tenth that of rubber is claimed) in hoses of the same capacity. It has been found that using bonded nylon hose more efficient refrigeration systems with smaller condensers, compressors and other components can be designed.

### 6.4.2 PACKAGING

Nylon of various types finds application in a variety of packaging items.

For securing rigid or semi-rigid containers nylon strip is now being used because of its lightness, strength and flexibility. It can be stapled like the steel strapping it replaces. Rigid bulk containers and drums for organic solids and the like are often lined with unlaminated nylon 6 or 66 film or sheets. Medical instruments may now be packed in sealed nylon film for subsequent thermal sterilisation in the pack.

Heat-stabilised grades of blown film are used for roast-in-the-bag food items because of the capability of retaining juice and of browning inside the pack during cooking.

Nylon 6, 10, 11 and 12 film is used for packing meat and fish products where, by using vacuum, close-to-form profiles are possible. Subsequent sealing of the bag prevents biological deterioration. Sterile medical disposable products are also packed in this way using nylon film. Thin nylon 11 and 12 film has also been used in continental Europe for the skins of smoked sausages.

Nylon 6 film laminated with polyethylene is used widely for flexible vacuum packing of perishable foodstuffs. This combination of materials ensures that the pack has high mechanical strength, impermeability to oxygen, odour resistance and oil (or fat) resistance. The PE layer also provides moisture proofness and heat sealability. The properties of the laminate are generally retained at both high and low temperature.

A major use of nylon-polyethylene laminate is in the manufacture of ready-for-use flexible pouches for bulk meat packs for deep-freeze storage. Other applications of pouches are for packing ground coffee, putty and mastics.

## Applications of polyamides

### 6.4.3 CHEMICALLY RESISTANT COMPONENTS

Nylon is used for a number of components in applications where chemical resistance is the prime requirement. Examples are carburettor floats, aerosol valves, pump seals and photographic processing equipment. In addition, nylon pipe and pipe fittings such as tees, elbows and couplings are now manufactured by some companies previously supplying these items in metal.

## 6.5 Miscellaneous applications

It is impossible to mention all the miscellaneous applications of polyamides, but a selection of proven utility, not covered in the classes discussed in previous sections, is listed in *Table 6.2*; this also

**Table 6.2** MISCELLANEOUS APPLICATIONS OF NYLON

| Industry/area of use | Fabrication method | Nylon type | Component |
|---|---|---|---|
| Agriculture | IM | 66 | Sprockets and dry-bearings for lawn-mowers |
| | E | | Wear pads for hay-balers |
| Automotive | IM and M | 66 | Steering-column bearings |
| | IM | 66 | Carburettor floats |
| | IM | 66 | Dashboard panels |
| | E | 66, 6 | Seat slides |
| | IM | | Door handles |
| | P and S | 66 | Distributor cams |
| | IM | 66 | Door and window stops |
| | E | 66 | Tyre chafer strips |
| Domestic and clothing | E | 6, 10, 11 | Meshes, food strainers, scourers |
| | IM | All | Cups and beakers |
| | IM | All | Syphon tops |
| | E | 66, 6, 11, 12 | Toothbrushes |
| | IM | 66, 6, 11, 12 | Combs |
| | IM | 66, 6, 11, 12 | Shoe-heel tips and studs |
| | IM | 66, 6, 11, 12 | Curtain fittings |
| | E | 66, 6, 11, 12 | Zip fasteners, curtain tape |
| | E | 66, 6, 11, 12 | Garment stiffeners |
| Furniture and office | IM | Various | Hinges, locks, curtain hangers, castors, furniture fasteners |
| | FBC | 12 | Chairs |
| | P and S | 66 | Addressograph rollers |
| | IM | 6, 66 | Pinions for typewriters |

## Applications of polyamides

**Table 6.2**—continued

| Industry/area of use | Fabrication method | Nylon type | Component |
|---|---|---|---|
| Marine | C, IM | 6, 12 | Propellers for small craft |
| | IM | 66 | Cleats for rigging |
| Medicine | E | Various | Syringes for injections |
| | E | Various | Catheters for transfusions |
| | IM | 66, 6, 6.10 | Spectacle frames |
| | BM | 66, 6, 6.10 | Containers for body fluids |
| Mining | C | 6 | Conveyor buckets |
| | C | 6 | Liners for coal washing equipment |
| Packaging | E | Various | Film and sheet for food packs (boil-in-bag) |
| | BM | 12 | Sheaths for sausages |
| Sports and toys | E | 11 | Tennis-racket strings |
| | E | 66, 6.10 | Fishing lines and nets |
| | E | Various | Ski-binding |
| | IM | Various | Fishing reels |
| | IM | 66 | Cigarette-holders and pipe mouthpieces |
| | IM | Various | Saddles |
| Textiles | C | 6 | Main drive-gears for knitting machines |
| | C, IM | 6, 66 | Dry bushes and bearings |
| Transport | E and M | 6, 66 | Fishplates for railway lines |
| | IM | 66 | Track insulators for railway lines |
| | IM, E | 66, 6 | Various aircraft components |

*Fabrication key*
| | |
|---|---|
| IM | Injection-moulded |
| E | Extruded |
| E and M | Extruded and machined |
| P and S | Pressed and sintered |
| C | Cast |
| BM | Blow-moulded |
| FBC | Fluidised-bed coating |
| M | Machined |

shows the fabrication method and nylon types used where these are known. The applications are conveniently collected under industry group or area of use. The automotive and domestic groups use by far the most of the miscellaneous nylon components, and business in this area is very large and continually expanding.

*Figure 6.5* shows the combination of two forms of nylon in a single assembly, a sintered nylon ink-roller supported in a nylon cassette forming a replaceable unit for the Friden print-out calculator.

## 6.6 Choice of production method

### 6.6.1 SMALL COMPONENTS

It is possible to produce certain classes of nylon component by a variety of methods. For example, gears may be produced by injection moulding, by machining from extrusions, as pressed and sintered finished components or as castings. For a given gear size, provided it is within the scope of the method, the choice of method is largely

Figure 6.5 Sintered nylon ink-roller in nylon cassette (courtesy Polypenco Ltd)

determined by the quantity of items required since this has a considerable effect on the overall cost. Injection moulding requires a high tooling cost, which must be spread over a large production run to reduce item cost. Casting, and pressing and sintering, incur far lower tool costs; for straight machining the tool costs are lower still. For short runs the latter methods are to be preferred. The choice of method is also dependent on the quality, the specified tolerances and the dimensional stability required of the part. Generally machining from stock shapes, either extrusions or castings, is used for close-tolerance parts.

## 6.6.2 LARGE COMPONENTS

Engineering components exceeding about 8 kg (17·5 lb) in weight are most economically produced using techniques other than injection moulding and sintering. There is also a practical limit to the size

Figure 6.6  Cast nylon gears in the paper industry (courtesy Polymer Corporation)

of extruded sections. In this region the methods of slush moulding or rotational or centrifugal castings from a melt may be used, but these techniques are mainly restricted to hollow bodies. Most of the nylons with good engineering properties may be handled using these techniques. The recently developed method of monomer casting using activated anionic polymerisation is, however, finding extensive use for both hollow and solid parts, particularly in the medium and heavy industries such as steel, shipbuilding and papermaking. Examples of these components are universal joints (slipper blocks) in drive spindles for steel rolls on rolling mills, stabilising fins for ships, and dryer gears used in Fourdrinier papermaking machines. This application is illustrated in *Figure 6.6*, which shows assembled cast nylon gears.

Using the casting technique many of the parts can be produced precisely to the specified size or will require only a minimum of machining for completion. The economics and applications of nylon castings are discussed by Summerhayes[3].

## 6.7 Applications of modified polyamides

Polyamide resins have wide application outside the strict fields of engineering plastics and textiles. They are used in formulations either alone, as the sole polymer, or in admixture with or adducted to other polymers, and often in the form of an organic solution, or organosol or water dispersion.

### 6.7.1 SOLUBLE NYLON FOR COATINGS AND ADHESIVES

Many copolymers of nylon are appreciably soluble in the lower alcohols and can be made up into solutions for adhesives, for forming moisture-absorptive and water-vapour-permeable coatings on fabric for wearing apparel, and like applications.

The molecule can also be chemically modified by reaction with formaldehyde in the presence of alcohol when alkoxy substitution occurs in the N-position. The resulting compound is quite soluble in the lower alcohols and, like the copolymers, can be used in coating or adhesive formulations.

Nylons derived from polymerised vegetable oils (section 3.5.5) are soluble in alcohol and alcohol-hydrocarbon mixtures, and in this form find wide use in forming thin flexible coatings. These may be made grease and water proof and can be heat-sealed. Solutions of nylons of this class are also used in adhesive formulations.

## 6.7.2 HOT-MELT ADHESIVES/MIXED ADHESIVES

Hot-melt adhesives usually consist of a blend of one or more vegetable-oil-based polyamide resins, together with modifying agents such as plasticisers and waxes, to confer strength at low temperature and reduce tackiness.

Polyamides of the vegetable-oil based class may also be combined with epoxy (and certain types of phenolic) resins to give thermosetting adhesives, where the epoxy resin is cured by the free amine groups in the polyamide.

## 6.7.3 INKS, ORGANOSOLS, WATER DISPERSIONS

Soluble polyamides, particularly those based on vegetable oils, have been used successfully in flexographic ink formulations required for plastic packaging material such as polyethylene, polyethylene terephthalate and cellulose acetate. This type of ink has some very exacting requirements, for example adhesion to substrate, gloss retention, water and fatty oil resistance, rub resistance and flexibility at low temperatures.

Polyamide resins may be dispersed in organic solvents as a colloidal suspension; this is particularly useful where the base resin has high melting point or low solubility. The colloidal particles are swollen but not dissolved, and the dispersion is thixotropic but fairly easy to handle and apply. On application to a surface the continuous phase evaporates leaving a discontinuous film of resin particles. This film may be fused by heating to 190°C or above.

Water dispersions of polyamides perform the same function as organosols in that resins, otherwise difficult to handle, are brought into a liquid form that can easily be applied to surfaces. The use of a water base eliminates the risks associated with solvent solutions, and cost reductions are effected. Various methods of preparing water dispersions of polyamides, and their properties, are described in the patent literature.

## 6.7.4 SPECIAL TYPES

Unusual applications for nylon are continually being reported in the patent and trade journals. Two taken at random are described briefly below.

## Applications of polyamides

### 6.7.4.1 Photosensitive polyamide resin

Claims have been made for polyamide compositions incorporating photosensitive compounds that effect polymerisation on exposure to light. This effect can be made the basis of a printing process using moulded plates of the polyamide composition[4].

### 6.7.4.2 Beer clarification

Anthocyanogens, which are important components of the non-biological haze that develops in some types of beer on storage, can be removed by filtration of the beer through purified nylon powder. The powder must be free of water-soluble material and must have suitable filtration characteristics. The nylon can be reactivated by suitable treatment. In a modern development of this process nylon powder has been superseded by polyvinyl pyrrolidone.

#### REFERENCES

1. (a) Verein Deutscher Ingenieure, *Bearings in thermoplastic materials without fillers*, VDI 2541 (1970)
   (b) British Standards Institution, draft standard produced by Technical Committee, MEE/119
2. Polypenco Ltd, *Bearing design* and *Gear design*, design manuals
3. Summerhayes, B., 'Nylon castings', *Engineering Materials and Design* (July 1970)
4. Time Inc., British Patent 862 276 (8.3.61)

# 7
# Characterisation, analysis and testing of polyamides

## 7.1 Features characterised

### 7.1.1 MOLECULAR WEIGHT, DISTRIBUTION AND GEOMETRY

Commercial polyamides, in common with other high polymers, comprise a mixture of molecules of various weights. If the manufacturing method produces linear molecules there results a distribution of chain length that may be measured and may be used to characterise the bulk polymer.

Since molecular chain length and distribution can be correlated with mechanical, thermal and other properties useful to the end-user, measurement and control of these characteristics is essential to obtain a standard quality in the production material.

The polyamide chain may be branched, cross-linked or cyclised or, as in the case of copolymers, the comonomer units may be segmented in regular or irregular sequence along the chain. These spatial characteristics can also be determined and controlled if uniform behaviour of the product is intended.

### 7.1.2 CRYSTALLINITY

The combination of stereo-regularity and secondary valence forces that exists in the majority of nylons leads to the partial crystallinity deduced from the X-ray diffraction pattern of samples in the solid state. Diffraction analysis may therefore be used to measure the percentage crystallinity of nylon polymers. As already discussed, this percentage relates closely to the mechanical properties.

A further advantage of X-ray diffraction methods is that they can be used to measure orientation in the bulk polymer. This is very useful in the study of fibres and filaments, which usually exist in a highly oriented state.

Two other methods commonly used to estimate crystallinity in polyamides are the infra-red and the density methods. All three methods mentioned are discussed later in this chapter.

### 7.1.3 FUNCTIONAL GROUPS AND MONOMERIC UNITS

The detection and estimation of the functional groups present in a polyamide help to identify the polyamide type. The end-groups of linear nylon homopolymers can be estimated provided the polymer can be dissolved in a suitable solvent. The total number of end-groups can be used to determine the number average molecular weight of the polyamide.

To identify commercial materials known to be of the polyamide type a combination of simple physical or chemical tests may be used, for example melting point, specific gravity and deflection temperature. If absolute confirmation is required the best method is to hydrolyse the polymer into its basic components, generally the monomer units, and separate and identify these.

## 7.2 Methods of characterisation

### 7.2.1 MOLECULAR STRUCTURE

#### 7.2.1.1 Viscometric methods

*(a) Solution viscosity*

The viscosity of a dilute solution of a polyamide can give information about the volume, shape and flexibility of the polyamide molecule and its interaction with the solvent. A discussion on the interpretation of viscometric measurements is beyond the scope of this book, but the principles have already been outlined in section 4.1.1, and a comprehensive detailed treatment is given by Peterlin[1].

Standard viscometric methods suitable for studying molecular structure and for quality control of the nylon manufacturing process are detailed in international standards, which are briefly described below.

(1) ISO (International Organisation for Standardisation), Recommendation R307, *Determination of the viscosity number of polyamides in dilute solution* (May 1963). The viscosity number has already been defined in *Table 4.1* and may be simply expressed as

$$\frac{\eta - \eta_o}{\eta_o C} \quad \text{which equals} \quad \frac{t - t_o}{t_o C}$$

where $\eta$ = absolute viscosity of the solution
$\eta_o$ = absolute viscosity of the solvent
$C$ = concentration (grams) of solute per ml of solution
$t$ = time of flow of solution
$t_o$ = time of flow of solvent

In practice formic acid or metacresol is the solvent and the polyamide solution is made up to a concentration of 0.005 g/ml. A viscometer of the suspended-level Ubbelohde type is brought into temperature equilibrium at 25°C before flow times are observed. The polyamide studied should give stable solutions and reproducible results for this method to be useful.

(2) ISO Recommendation R600, *Determination of the viscosity ratio of polyamides in concentrated solution* (August 1967). This method is substantially the same as that described in ASTM Standard D789. The viscosity ratio, hitherto termed relative viscosity, was expressed in *Table 4.1* as $\eta/\eta_0$ and may be calculated from the expression

$$\frac{K}{K_o} \cdot \frac{T}{T_o} \cdot \frac{d}{d_o}$$

where $K$ = constant of viscometer used for measuring viscosity of solution
$K_o$ = constant of viscometer used for measuring viscosity of solvent, in this case formic acid
$T$ = time of flow of solution, in seconds
$T_o$ = time of flow of solvent, in seconds
$d$ = relative density of solution
$d_o$ = relative density of solvent

In this method the solvent is specified as a 90 per cent w/w aqueous solution of pure formic acid and the polyamide concentration is 8.4 per cent polyamide by weight in the solution. An Ostwald-type viscometer is used and flows are measured at 25°C.

In using both the above viscometric methods the presence of monomer in the polymer may considerably affect the viscometric behaviour; it is therefore usual to establish if monomer is present and, if found, to estimate it. Fillers and additives, if known to affect the viscosity, must be removed from samples under test.

Temperature has a marked influence on the viscosity of solutions and for precise measurements it should be controlled at $\pm 0.01°C$. For control purposes $\pm 0.1°C$ may be tolerated. The dilute-solution method of ISO R307 is used primarily for the study of molecular structure; the concentrated-solution method of R600 is used mainly for quality control during manufacture of polyamides.

*(b)   Melt viscosity*

Like solution viscosity the melt viscosity of polyamides can serve to yield information regarding molecular weight, and a relationship has already been discussed (section 4.1.1). Strictly this relationship holds good only for low or medium average molecular weights or in cases where the rate of shear is sufficiently low not to affect the viscosity. In these cases the melt behaves as a Newtonian fluid. For all other cases it is usual to express the measurement as an apparent melt viscosity. This point is illustrated in *Figure 7.1*, which

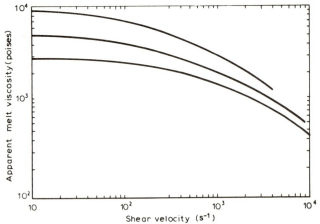

Figure 7.1   Apparent melt viscosity vs shear velocity for nylon 6 ($\eta_{rel} \approx 2.8$)

shows the relationship between shear rate and melt viscosity of nylon 6[2].

As expected the viscosity of a polyamide melt decreases with temperature, and the following characteristic relationship is obeyed:

$$\log \eta = A + B/T \tag{7.1}$$

where $T$ = absolute temperature of melt (K)
$A, B$ = constants

Two other factors that influence the viscosity of some polyamide melts may have to be taken into account: the monomer content, which tends to lower the melt viscosity, and the presence of substantial amounts of inert filler, which tends to increase the melt viscosity.

In considering the very many methods described in the literature that purport to measure melt viscosity, the methods that give true shear-independent viscosities measured in absolute units (poises, stokes or derivatives) should be distinguished from those that give apparent viscosities at some arbitrary chosen shear rate. Only the former methods can give useful information on molecular parameters and characterisation. The methods in the latter category, however, find widespread favour in determining the flow properties of polymer melts for the purpose of processing control. This is discussed more fully in section 7.4.1.

### 7.2.1.2 Osmometry

Number-average molecular weights of polyamides can be obtained from osmotic pressure measurements of dilute solutions of the polymer using the van 'tHoff expression:

$$(\Pi/C)_o = RT/M \tag{7.2}$$

where $\Pi$ = osmotic pressure
$C$ = concentration of solute
$R$ = gas constant
$T$ = absolute temperature
$M$ = molecular weight

The term $(\Pi/C)_o$ is obtained from measurements of $\Pi$ at various concentrations and extrapolation to zero.

The most important experimental feature in osmotic measurement is the membrane, whose permeability must be large enough for equilibrium to be attained in a reasonable time but not too large to allow lower molecular weight materials to pass. The membrane material must also be compatible with the solvent.

The osmotic method of determining molecular weight of polyamides is most accurate when applied to material fractionated into narrower molecular weight bands.

### 7.2.1.3 Light scattering

Light-scattering measurements on polymer solutions can yield data on which the weight average molecular weight, $\overline{M}_w$, of the polymer system can be calculated. Since only geometric measurements are made and basic physical constants are used in calculation the method is absolute and simple in principle. In practice a great deal of experimental skill is needed to ensure the best results.

### 7.2.1.4 End-group analysis

The number-average molecular weight, $\overline{M}_n$, of a commercial polyamide, provided it is soluble in a suitable solvent, is most frequently determined by the method of end-group analysis. This method depends on the fact that each chain is terminated by an amine, a carboxyl, or one of each type of functional group. In the case of end-group stabilised polyamides, one of the functional groups mentioned is replaced by a chain stopper; for example, acetic acid is commonly added in the polymerising mix to react with amine endgroups and effectively prevent further condensation at that end.

To determine molecular weight, all end-groups must be estimated and expressed usually in terms of equivalents per $10^6$ grams of polymer. Since there are two ends to every linear polyamide molecule the number-average molecular weight can then be calculated from the expression:

$$\overline{M}_n = \frac{2 \times 10^6}{\text{Equivalent (amine + carbonyl + stabiliser) ends}/10^6 \text{ g}}$$

For end-group analysis of copolymers or homopolymers of low molecular weight, the sample may be dissolved in ethanol-water mixtures and titrated potentiometrically using, separately, dilute sodium hydroxide and hydrochloric acid to estimate respectively carboxyl end-groups and amine end-groups[3]. Most of the commercial polyamides require a stronger solvent such as m-cresol for complete solution before titration is possible.

Conductometric titration has been used for determination of amine end-groups in unstabilised nylon using a phenol-ethanol solvent for the polymer[3]. The precision of the conductometric method is stated to be less than for the potentiometric method, due possibly to the pronounced effect of small amount of impurities.

Carboxyl end-groups in polyamides can be determined by titration of the benzyl alcohol solution of the polymer using dilute standard sodium hydroxide[3].

For acetic acid stabilised polyamides the acetyl end-groups can be determined by first hydrolysing the polymer, followed by separation and estimation of the acetic acid formed in the reaction.

### 7.2.1.5 Fractionation

The distribution of chain-lengths (and hence molecular weight distribution) is an important characteristic in a commercial polyamide since changes in the distribution affect the processing, particularly flow properties as well as the final properties of the polymer. The theoretical aspects of the development of distributions of chain-lengths in condensation polymers have been discussed thoroughly by Flory[4] and a review of fractionation methods given by Hall[5].

It is inappropriate here to discuss the merits of the methods that have been used for fractionating polyamides. Most of these are elaborate and time-consuming, and the number of fractions determined and the sharpness of each is largely related to the time-scale of the process. Preparative fractionation, where physical separation of the individual fractions is carried out and followed by separate molecular weight determinations, should be distinguished from analytical fractionation, where the molecular weight distribution can be determined without isolation of the fractions. In the former class it is important to mention the new rapid fractionation technique of gel-permeation chromatography. This separation method uses column chromatography, in which the stationary phase is a hetero-porous solvent-swollen polymer network varying in permeability over several orders of magnitude. As the polymer solution passes through the gel column the polymer molecules diffuse into areas of the gel appropriate to their size, the smaller molecules permeating more completely and hence spending more time in the column than the larger molecules, which pass through more quickly. Differential refractometry can be applied to the solution eluting through the column, to detect the course of the molecular separation and serve as a basis for collection of fractions.

It has been shown that the elution volume of a polymer is proportional to the logarithm of its molecular weight and, from the GPC curves obtained from the chromatogram, the molecular weight distribution curves can be obtained by calibration and mathematical analysis.

Using GPC, polymeric systems can now be separated into fractions in far less time and using a much smaller quantity of material than hitherto possible by other methods.

### 7.2.2  CRYSTALLINITY

#### 7.2.2.1  X-ray diffraction

X-ray diffraction diagrams of partially crystalline polymers such as polyamides show peaks of high intensity corresponding to the crystalline regions, and bands of low intensity corresponding to the amorphous regions. Integration of the areas under the two portions of the curve gives an approximate measure of the crystalline/amorphous ratio, but the precision of the method is limited by the ability to resolve the crystalline and amorphous patterns.

The percentage crystallinity of the common commercial nylons is usually between 40 and 70 per cent.

#### 7.2.2.2  Infra-red method

The infra-red spectrum of nylon contains a number of bands that vary in intensity either with the crystallinity or with the amorphous content of the polymer. For example, for nylon 66, the 10.68 μm band can be used to follow the crystallinity variation and the 8.78 μm band the amorphous. If samples are prepared varying in crystallinity over a wide range, a plot of the measured density against absorption intensity of the crystalline and amorphous bands shows a linear relationship; extrapolation to zero intensity then enables the fully crystalline and amorphous densities to be determined. The amorphous density of the polymer can also be determined by direct measurement of the fully quenched material, but this depends on the assumption, not always valid, that no crystalline material is present in the quenched sample. Again, the fully crystalline density can be calculated from unit cell constants obtained from X-ray data, although with less precision than can be obtained by the infra-red method.

Starkweather and Moynihan[6] used the infra-red method to obtain crystalline and amorphous densities of the common nylons. Their results are included, with those from other investigators, in *Table 7.1*.

Table 7.1 CRYSTALLINE AND AMORPHOUS DENSITIES OF COMMON NYLON TYPES

| Nylon type | Crystalline density (g/cm$^3$) | Amorphous density (g/cm$^3$) |
|---|---|---|
| 66 | 1.22* | 1.07* |
|  | 1.24 | 1.09 |
| 6.10 | 1.15* | 1.04* |
|  | 1.16 | 1.05 |
| 6 | 1.23 | 1.09 |

*Starkweather and Moynihan's data

### 7.2.2.3 Density method

The density method for determining the degree of crystallinity of a nylon whose type is known can be used, provided the fully crystalline and fully amorphous densities of the type have already been established from X-ray diffraction or infra-red data as described in the previous sections. If the assumption is made that the decrease in specific volume from that of the completely amorphous material is proportional to the amount of crystallinity, then the volume fraction of crystallinity, $x$, can be found from the relationship:

$$x = (V_a - V)/(V_a - V_c) \tag{7.3}$$

where  $V_a$ = specific volume of purely amorphous material
$V_c$ = specific volume of purely crystalline material
$V$ = specific volume of sample

The density method of determining crystallinity is useful for routine measurements of crystallinity since it is simple and fast. Care must be taken to ensure the samples used are free from voids.

## 7.3 Identification and analysis

### 7.3.1 IDENTIFICATION

If the sample to be examined is known to be a commercial product, and a polyamide, the determination of melting point and specific gravity is usually sufficient to identify the type, if reference is made to tables listing these properties.

#### 7.3.1.1 Infra-red spectra

Examination of the infra-red spectra of the samples allows identification of the material as a polyamide, as well as determining the specific type. Again reference is made to known standards for comparison of the absorption bands. Difficulty has been found in identifying copolymers by this means and for these other methods are usually employed.

Nylons are usually characterised by the bands in the 670–5000 $cm^{-1}$ region arising from the peptide linkage, for instance the 3050 $cm^{-1}$ band due to the N—H stretching vibration of the secondary amide, the 1650 $cm^{-1}$ band due to the carbonyl stretching frequency and the 1550 $cm^{-1}$ band due to the N—H deformational vibration. These bands are also present in other compounds containing the secondary amide groups, for example the proteins, but the other bands in the nylon spectrum are specific to the type, and permit identification.

Since the intensity, and to some extent the location, of the absorption bands depend upon the degree of crystallinity of the sample it is advisable to anneal the sample to reduce crystallinity effects before examination.

Commercial polyamides usually contain organic stabilisers, antioxidants, and other improver additives; to avoid confusion with the base polymer spectra these additives are preferably removed before examination.

Samples suitable for infra-red examination may be in the form of films prepared from a solvent paste, or cast from a solution. For insoluble polyamides it is also possible to use hot pressed films or to cut very thin sections (2–3 µm) using a suitable microtome. In all these cases it is important to control the sample thickness within quite fine limits.

A recent technique, which has the advantage of time-saving in sample preparation and can be classed as non-destructive in that no processing is required in preparation, is that of attenuated total reflectance. In this method an interface is made between the plastic sample and one of very much higher refractive index (RI). The infra-red beam is directed on to the interface approaching from the high-RI side at about 45°, when a large amount of the energy is reflected at the interface. The beam traverses the interface and penetrates the low-RI material to a depth of several micrometres. If several reflections are arranged, sufficient attenuation of the beam occurs to yield a good absorption spectrum. The high-RI prism or component commonly used is a mixed thallium bromide/iodide crystal of RI 2.6. Since the only requirement is good optical contact between the sample material and the high-RI prism, preparation of samples is minimal and the technique is suitable for insoluble polyamides in many forms; it has partially replaced the use of mulls, pellets and hot pressings hitherto popular for insoluble materials.

The use of organic solutions of polyamides is limited because most solvents have their own characteristic absorption bands in the infrared, often located near those of the polyamide, and interference can result. Selected peaks may be examined if the solvent is known to show low absorption in that region, and direct compensation is possible if double-beam instruments are used, with the polyamide solution and the pure solvent in separate beams.

Pyrolysis of the sample followed by infra-red examination of the pyrolysate has been used as a technique by Harms[7] for nylon 66. The method is useful if normal infra-red examination cannot be carried out because of fillers and other additives. Full documentation is required on the infra-red pyrolysates of known polymers so that comparison can be made.

It has been found difficult to identify nylon copolymers from their infra-red spectra and in these cases analysis of the hydrolysates has been found satisfactory.

Methods for the analysis of nylon and related polymers using infra-red spectra and from examination of hydrolysates are described by Haslam and Willis[8].

### 7.3.1.2 Ultra-violet spectra

Ultra-violet (UV) absorption involves electronic transitions in the molecule associated, in the main, with the $\pi$ electrons of conjugated systems; the absence of the latter in polyamides implies that they are normally UV transparent. Where UV absorption is found it is usually due to impurities, fillers or other additives. Thermal degradation of nylon 66 has also been shown to lead to development of an absorption in the polymer at 290 nm and this observation has been utilised to follow the course of degradation in this nylon[9].

### 7.3.1.3 Chromatographic methods

Paper partition chromatography has been used to separate and identify a mixture of the common polyamides nylons 6, 66 and 6.10. This technique was developed and is described by Ayres[10]. The samples are dissolved in 90 per cent acid, spotted on Whatman No. 54 filter paper, and developed with 88 per cent formic acid. A coloured front shows up on the paper applying Solacet Fast Blue 2 BS in acetic acid. The $R_F$ values reported were 0.8 for nylon 6, 0.4 for nylon 66, and 0.3 for nylon 6.10. The method can be used to distinguish between copolymers and mixtures of homopolymers.

Pyrolysis followed by gas chromatography has been used in routine identification of commercial polyamides[11]. This is a 'fingerprinting' method in that the pyrograms of the unknowns are compared with those of known polyamides pyrolysed under identical conditions. To minimise variations caused by small differences in experimental conditions a single peak may be selected as a reference, and retention times of other peaks expressed as a percentage.

### 7.3.1.4 Miscellaneous methods

*(a) Solubility*

The solubility of many of the nylons in 90 per cent formic acid can be utilised in identifying the common nylons, 6, 66 and 6.10. Among the common high polymers, cellulose esters are the only other group soluble in this acid, but the solubility of the ester in acetone also serves to distinguish the two polymeric types since nylons are insoluble in acetone.

Once the sample has been identified as a nylon it is possible to distinguish between nylon 6, 66 and 6.10 using the simple solubility scheme shown below[3].

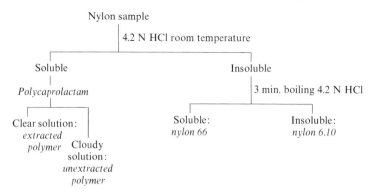

Polycaprolactam of the cast nylon type and branched or cross-linked nylons cannot be identified by simple solubility methods. They are insoluble but may be swollen in formic acid and are best identified by analysis of their hydrolysates.

### (b) Specific gravity

If specific gravity is used as an identifying feature an accuracy to three significant figures is required. A convenient method of obtaining this is to use the concentration gradient density tube. Here by controlled mixing of two miscible solvents (one heavy, the other light) and uniform filling of a graduated tube, a smooth gradient of density over the filled length of the tube can be achieved. If the tube is put in a thermostatted bath, vertical mixing by convection is minimised and the gradient remains stable for many months. Samples of unknown density placed in the tube take up a rest position at the level of equivalent density. The gradient is usually determined first before use by observing the position, relative to the tube's fixed graduations, of floating glass markers of known density and plotting a calibration curve. The method of preparing a concentration gradient density tube is described in BS 3715. Carbon tetrachloride and xylene have been used to obtain a gradient of 1.10 to 1.60 g/ml, which is suitable for most of the unfilled nylons.

### 7.3.2 ANALYSIS OF POLYAMIDES

The identification methods already described are suitable if the base polymer is a single type and if fillers and additives do not interfere. The one exception is the paper chromatographic method using formic acid, where mixed homopolymers and copolymers can be separated and identified.

### 7.3.2.1 Hydrolysis and chromatography

The more popular method for qualitative analysis of polyamide resins is to hydrolyse the polymer into its basic constituents and identify these by chromatography. This method has been found particularly useful for copolyamides.

In one method[12] the samples are hydrolysed by refluxing in 6 N hydrochloric acid. The amine hydrochlorides are esterified and ether extracted and then chromatographed on a 5 per cent diethylene glycol adipate packed 2 m (6 ft) column. Another portion of the hydrolysate is neutralised and the free amines extracted with $n$-butyl alcohol and chromatographed on a 2 m (6 ft), 6 mm ($\frac{1}{4}$ in) diameter column packed with 1 per cent of Apiezon L or glass beads. Nylon copolymers 66/6.10 and 66/6.10/6.12 have been successfully analysed by this method. As aminocaproic acid is not resolved the amount of nylon 6, if known to be present in the copolymer, is obtained by difference.

Very detailed methods of analysis of hydrolysates of the common nylons 66, 6, 6.10 and 11, and copolymers involving these, are given in papers by Haslam and Clasper[13]. After preliminary removal of plasticiser, for which methods are given, 50 per cent v/v hydrochloric acid was used to hydrolyse the polyamide sample. Ether extraction of the hydrolysate was used to remove the acids, and the base hydrochlorides were recovered from the aqueous solution remaining after ether extraction. Methods for the separation and identification of individual components in the two main fractions are detailed in the papers.

For analysis of small (milligram) quantities of polyamide samples Haslam, Clasper and Moodey[14] developed a paper chromatographic method. The ethanolic solution of the dried acid-free hydrolysate was used for spotting, and ninhydrin and methyl red borate used to locate the positions of the base and acid constituents, respectively, on the developed chromatogram.

Haslam and Udris[15] further developed a ring paper chromatographic method for analysing the hydrolysates using reagents similar to those described previously for dissolving the hydrolysate, developing, and locating the positions of constituents. This variation of the chromatographic technique produces rings on the paper with $R_F$ values characteristic of the constituents of the hydrolysis, and hence of the polyamide in the sample.

### 7.3.2.2 Constituent analysis

In addition to the base polyamide, nylon plastics materials contain a number of constituents of which some take part in the basic chemistry of polymer formation and are present in small amounts in equilibrium with the polymer—for example, water and monomer. Other constituents are deliberately added to perform a specific function and are reactive with the polymer, if at all, only to a minor degree. In this second category are plasticisers, antioxidants and other stabilisers, and fillers, including colourants. Degradation products arising from processing or service use may also be present.

When it is known that many analytical processes because of their severity can destroy or modify the constituents that are being estimated, a complete constituent analysis of a polyamide plastic can be a formidable task. Fortunately, well established methods are available for many of the common constituents and their analysis is outlined below.

*(a) Moisture*

The most popular method for determining moisture in moulding granules or fabrications therefrom is Method B of ISO Recommendation R960. ASTM D 789 describes a similar method. The polyamide sample is melted under vacuum at approximately 30°C above its melting point, and the amount of water collected in a cold trap is determined by the Karl Fischer method. For low water contents a correction should be made for water of condensation produced during the drying process.

ISO R960 Method A is also applicable to nylon granules. Here the moisture is extracted using anhydrous methanol and determined in the extract by the Karl Fischer method. ISO R960 Method C can be used for nylon powders. The sample is dissolved in boiling meta-cresol and moisture in solution determined by the Karl Fischer method.

## (b) Monomer

Nylon polymers of the AA/BB type contain small quantities of monomer in equilibrium with the polymer. The presence of this monomer can modify the properties of the polymer in a similar way, but to a much less degree than moisture, by acting as a plasticiser.

Polymers of the AB type produced from lactams generally contain higher proportions of monomer in equilibrium with the polymer, and their effect is proportionally greater than for the AA/BB type polymers. It is not always possible nor often necessary to separate monomer from low molecular weight polymer (dimer, trimer, etc.) in determination of 'monomer content'. The effect of all such molecules on properties is nearly the same. ISO Recommendation 512 details methods for extraction and estimation of monomer and low molecular weight polymer in polyamide resins. Separate methods are given for nylon 6 and 66, 6.10 and 11 as two groups. Results are adjusted to account for the water content, estimated separately, and if possible the plasticiser if this is present and can be estimated.

Ongemach and Moody give a method for determining residual caprolactam in nylon 6 that involves water extraction followed by chromatographing the aqueous solution[16].

Haslam[8] describes a further method for determining free caprolactam in nylon 6 where the infra-red spectrum of the sample is compared with that of a monomer free sample. Specific absorption bands due to the monomer are observed and measured at 11.5 μm.

## (c) Inert fillers

Inert fillers are those that are not attacked by dilute (approximately 6 N) hydrochloric acid; they include all types of carbon, molybdenum disulphide, E-glass and certain colouring pigments. If these fillers are known to be present they can be readily estimated by hydrolysing the nylon sample, filtering off the unattacked filler, drying and weighing.

For certain fillers such as titanium dioxide, known to be stable at furnace temperatures, estimation can be made using a dry ashing technique and determining the residue weight. Trace metal determination on nominally unfilled polyamide may also be carried out using an ashing technique followed by colorimetric analysis.

*(d) Organic additives*

The organic additives under this heading include plasticisers and anti-oxidants. The estimation of such compounds in nylons can be a difficult task. This is because many of these compounds, particularly anti-oxidants, have a high chemical reactivity and low stability; they are usually in low concentration in the polymer and therefore difficult to isolate, and they are intimately mixed in the polymer matrix. The presence of several such additives in one formulation can be an added difficulty and there is also the possibility of the confusing effect of polymer thermal degradation products present in the sample.

Plasticisers in nylons are best determined by infra-red analysis or gas phase chromatography, after extraction from the polymer. Haslam[8] describes a method of extraction of *p*-toluene sulphonamide plasticiser from a nylon 66/6.10/6 copolymer.

Pyrolysis of the polymer followed by gas phase chromatography of the pyrolysate has been used, advantage being taken of the fact that the pyrolysis products of the polymer itself are usually much lighter and elute more readily than those of the plasticisers and can be separated in the chromatogram.

In the case of anti-oxidants, if the presence of specific compounds in the nylon sample is known, direct spectroscopy can often be used to estimate the amount present. Solvent extraction is preferred in the case of amine anti-oxidants, when the extract will usually show some absorbence in the UV proportional to the amount of additive present. *Nonox B* (ICI trademark) for instance gives an absorption peak at 290 nm.

Phenol-type anti-oxidants may be coupled to form coloured compounds that can be estimated spectroscopically.

Details of the methods mentioned above and many others are described by Crompton[17].

## 7.4 Testing for quality, type and design

In order that the end-user can rely on the quality and performance of the polyamide he is using, the material must be subjected to a comprehensive series of tests at various stages along the route from polymerisation of the resin to the final products. The requirements are usually detailed in specifications, which may take the form of Quality Control Specifications or Material Type Specifications.

Basic information is also required by designers on the time-dependent, rate-of-loading-dependent and environment-dependent properties so that the lifespan of the design can be forecast. These three forms of testing will be discussed for polyamides.

### 7.4.1 QUALITY CONTROL TESTING

In the initial polymerisation, whether this be a batch or continuous method, in-process control must be adequate to ensure that the polymer has properties within the specified tolerance bands. Batches of resins must be related to a complete cycle of events that are unique in that the following cycle produces a discrete quantity of resin, differing slightly in properties due to slight changes in the process variables. In the case of batch production each batch must be sampled (and if necessary batches blended to form larger units that are resampled) to ensure property values fall within specified limits.

For continuous production, batches may be defined as the quantity of resin produced in a given period of time. Again, blending to form larger units, then retesting, is common practice. Rapid but accurate tests such as those used to determine solution viscosity, melt viscosity and moisture content are frequently used to control quality.

The fabricator who converts the resin to stock shapes or moulded parts expects to use material of the correct nylon type and of specified quality. His own conversion is subject to similar quality criteria, and the end-user expects to receive shapes or parts within the specification band agreed between the fabricator and himself. Quality tests in this case are, commonly, tensile strength and elongation, hardness, and relative viscosity of the polymer solution; these are all rapid and reliable. Frequently a customer will specify that a particular test (e.g. dimensional stability), important for his end-use, be carried out and he will lay down the limits within which the value of a particular property must fall. The test method used must take account of directional effects in the fabrication tested.

### 7.4.2 TYPE TESTING

Commercial polyamides are available in many types, distinguished by the basic repeat unit in the polymer. For each type grades exist

## Characterisation, analysis and testing

distinguished by molecular weight average, molecular weight spread, additive type, or other features.

Fabricators and end-users are faced with a very wide choice of material properties, and the object of type testing and type specifications is to ensure that the definitive properties of the type are specified and are met by the sample tested.

National standards exist that define commercial types and grades of nylons for specific purposes. Examples are shown in *Table 7.2*.

Table 7.2 STANDARD DEFINING TYPES AND GRADES OF NYLON

| Specification | Origin | Title | No. of types specified | No. of grades specified |
|---|---|---|---|---|
| ASTM D789 | US | Nylon injection moulding and extrusion materials | 7 | 15 |
| L-P-410a (Federal Spec.) | US | Plastic, polyamide rigid: rods, tubes, flats, moulded and cast parts | 4 | 4 in each type |
| R1874 (Recommendation) | ISO | Specification for polyamide homopolymers | 5 | 39 |

Raw-material suppliers can generally supply material to meet the types and grades in the above table, and in addition will supply other grades of their own designed for special purposes. Based on type specifications material suppliers often issue, for sales purposes, product data sheets or brochures listing typical properties.

Type specifications must detail the test methods for the properties or performance specified, and the reader is referred to the documents in *Table 7.2* or to ISO, ASTM or BS specifications for the methods commonly used. A recent monograph[18] describes in detail the whole subject of plastics test methods, with one chapter devoted to standards and specifications. Apart from the ASTM Standards, particularly those in Parts 35 and 36, the writer has found US Federal Test Method Standards widely used throughout Europe. The use of German DIN Standards tends to be restricted to continental Europe, just as British Standards are mainly used in the UK. With the entry of the UK to the EEC, it is likely that in the future EEC Standards based on ISO standards will be used.

## 7.4.3 TESTING FOR DESIGN

Until recently and with very few exceptions the availability of data for designing in polyamides has been restricted to the publications of the main raw-material suppliers or to special articles in trade or scientific journals.

The large engineering firms and Government ministries using polyamides for civil or military purposes often carry out their own testing for design. Where designs are critical, as in the aircraft and some aspects of automobile manufacture, an elaborate quality-assurance organisation is employed to ensure that the quality and type of material used in particular design are as specified by the designer.

The British Standards Institution, in BS 4618, *Recommendations for the presentation of plastics design data*, has given guidelines for the presentation of data on the properties of plastics materials for use in design engineering. The standard points out the inadequacy for design purposes of 'single-point' values of properties under closely defined test conditions such as are given in quality control tests; recommendations are therefore given for the collection of multi-point data related to changes in a property with time, temperature, humidity and other service conditions.

The availability of reliable design data for the many types and grades of polyamides now on the market must further consolidate their use as reliable engineering materials.

### REFERENCES

1. Peterlin, A., in 'Die Makromolekul in Losungen', Vol. II of *Die Physik der Hochpolymeren* (Stuart, H. A., ed.), 305–309, Springer, Berlin (1953)
2. Bernhardt, E. C., *Processing of thermoplastic materials*, Reinhold Corp., New York (1959)
3. Kline, G. M. (ed.), *Analytical chemistry of polymers: High polymers*, Vol. XII Part I, 277–290, Interscience (1959)
4. Flory, P. J., *Principles of polymer chemistry*, Cornell University Press (1953)
5. Hall, R. W., in *Techniques of polymer characterisation*, Allen, P. W. (ed.), Academic Press, New York (1959)
6. Starkweather, H. W. and Moynihan, R. E., *J. Polymer Sci.*, **22**, 363 (1965)
7. Harms, D. W., *Analytical Chemistry*, **25**, 1140 (1953)
8. Haslam, J., Willis, H. A. and Squirrell, D. C. M., *Identification and analysis of plastics*, 2nd edn. 292–300 and 312–324, Newnes–Butterworths (1972)
9. Matthews, J. M., *Textile fibers*, 6th edn, Chapter 20, Wiley, New York (1954)
10. Ayres, C. W., *Analyst*, **78**, 382 (1953)
11. Cox, B. C. and Ellis, B., *Analytical Chemistry*, **36**, 90 (1964)
12. Lehmann, F. A. and Brauer, G. M., *Analytical Chemistry*, **33**, 673 (1961)

13  Clasper, M. and Haslam, J., *Analyst*, **74**, 224 (1949); Haslam, J. and Clasper, M., *Analyst*, **76**, 33 (1951)
14  Haslam, J., Clasper, M. and Moodey, E. F., *Analyst*, **82**, 101 (1957)
15  Haslam, J. and Udris, J., *Analyst*, **84**, 656 (1959)
16  Ongemach, G. C. and Moody, A. C., *Analytical Chemistry*, **39**, 1005 (1967)
17  Crompton, T. R., *Chemical analysis of additives in plastics*, Pergamon (1971)
18  Ives, G. C., Mead, J. A. and Riley, M. M., *Handbook of plastics test methods*, Plastics Institute Monograph, Newnes–Butterworths (1971)

# Index

Acrylonitrile dimerisation process, 14
Adhesives, use of nylon for, 202–203
Adipic acid, 11, 12–13
Ammonium sulphate, 11, 18
Analysis,
　chromatographic methods of, 211–212, 216, 218–219
　constituent, 219
　end-group, 210–211
　infra-red spectra, 214–215
　moisture-content, 219
　monomer-content, 220
　ultra-violet spectra, 215–216
Anionic polymerisation, *see* Polymerisation
Anisomorphous, *see* Copolyamides
Annealing, 159
Applications of nylon,
　chemical, 195–198
　coatings and adhesives, 202–203
　electrical, 194–195
　friction and wear, 192
　gears and bearings, 190–192
　general engineering, 189
　housings and casings, 193–194
　inks, organosols, water dispersions, 203
　miscellaneous, 198–200
　packaging, 197
　rotating components, 193
　special, 203–204
　telecommunications, 195

Barrel heating, 165
Bearing properties, 114, 130
　effect of fillers on, 116–117
　evaluation of, 114–116
Bearings, nylon, 190–192
Beckmann rearrangement, 18

Blanking of nylon strip, 186
Butadiene, 12, 13

Cable-sheathing extrusion, 170–171
Caprolactam, commercial processes, 14–24
Carothers, Wallace N., 2, 3, 4
Cast nylon, 173–178
　component mixing, 174–176
　intermediates, 174
　manufacturing techniques, 176
　mould design, 177
　post-moulding treatments, 177
　special formulations, 177–178
Castor oil, 11
Charpy test, 84
Chemical applications of nylons, 195–198
Chemicals, effects of, 62–69
　inorganic, 65–66
　oxidising agents, 67
　solvents, 68
Chromatographic analysis, 211–212, 216, 218–219
Coatings, use of nylon for, 202
　powder types, 178–181
Compatibility, chemical, 64–69
Compressive properties, 80–81
Copolyamides, 44–46
Creep, 89–96
　derivative curves, 92
　effect of variables on, 94
　recovery from, 97–98
Crystalline/amorphous ratio, use in analysis, 213
Crystallinity, 59–62, 205–206, 212–213
Cyclic oligomers, 46–48
Cyclohexane oxidation, 16
Cyclohexanone, feedstock, 16

227

Damping, mechanical, 87–88
Deflection temperature, 136–137
Degradation of nylons, 69–78
  biological, 78
  effect of fillers on, 77–78
  high-energy radiation, 78
  light, 75
  thermal, 71–74
Dielectric loss, 139–140
Dielectric strength, 140–141
Diol process for HMD, 14
Direct conversion (cast nylon), 173–178
Dodecalactam, *see* Laurinlactam
DSM sulphuric acid recycle process, 21
Du Pont, 2, 3

Electrical applications of nylons, 194–195
Electrical properties, 137–143
  effect of fillers, 143
  effect of frequency, 141–143
Electrostatic coating, 181
End-group analysis, 210–211
Engel process, 9
Enthalpy of plastic materials, 147
Environmental effects, 69–78
Extrusion, 160–172
  change of material, 166
  fast-running screws, 163–164
  film, 171
  melt cooling, 166
  monofilament, 171–172
  processing conditions, 164–166
  process variables, 162
  rod and plate, 167–168
  screw design, 161–162
  tube, 168
  twin-screw, 164
  vacuum extraction, 163
  wire coating/cable sheathing, 170–171

Fade-O-Meter, 76
Fatigue,
  dynamic, 100–102
  static, 98–100
Feedstock,
  cyclohexane, 11
  phenol, 15, 16
Fillers,
  effect on bearing properties, 116–117
  effect on degradation, 77–78
  effect on electrical properties, 143

Fillers *continued*
  effect on thermal properties, 135
Film extrusion, 171
Finishing operations, 182–186
Flame spraying, 180–181
Flexural properties, 82
Fluidised-bed coating, 179–180
Fractionation, 211–212
Friction applications for nylon components, 192
Friction properties of nylon,
  effect of additives, 107
  effect of moisture, 129
  effect of temperature, 107
  theories, 104, 105
Fringed micelle model, 59
Functionality, 4

Gears, nylon, 190–192
Gel-permeation chromatography, 211–212

Hardness, measurement of, 82–84
Heat of polymerisation, lactams, 35
Hexamethylene diamine, 12
Hill, J. W., 2
Housings, nylon, 193–194
Hüls, laurinlactam production, 24
Hydrogen bonding in polyamides, 60
Hydrolysis, 218–219
Hydrolytic polymerisation, 53
Hydroxylamine for caprolactam production, 17–18

I. G. Farben, 3
Impact properties, 84–87
Infra-red spectra, use in characterisation and analysis, 212–213, 214–215
Injection moulding, 145–160
  characteristics of nylon types, 148–149
  mould design, 153–155
  moulding cycle, 151
  moulding defects, 155–158
  moulding machines, 150–153
  moulding materials, 149–150
Inks, use of nylons in, 203
Isomorphous, *see* Copolyamides
Izod test, 84

KA oil, 11, 14, 16
Kanebo process, 21

# Index

$L/D$ ratio, 161–162, 163
Lactone process for caprolactam, 20
Laurinlactam, 24
Light-scattering, 210
Long-term properties, 89–102

Machining, 182–186
Mechanical applications of nylons, 189–194
Melt spinning, 172–173
Moisture absorption,
  effect on dimensions, 122–125
  effect on electrical properties, 137–139
  effect on friction, wear and bearing properties, 129–130
  effect on long-term properties, 128–129
  effect on short-term properties, 125–128
  equilibrium, 120–121
  graphical presentation, 121–122
  kinetics, 117–119
Moisture content, analysis of, 219
Moisturising, 159–160
Molecular still, 4
Molecular structure, 55–59, 206–212
Molecular weight,
  averages, 55–56
  control, 52
  distribution, 57–59
  measurement of, 210, 211
  relationship with properties, 55
Monofilament extrusion, 171–172
Monomer, 26–27
Monomer content, analysis of, 220
Morphology, 59–62
Mould design, 153–158
Mould gates, 153, 154
Mould venting, 155
Moulding characteristics of nylon, 146–148

Nucleating agents, 150
Nylons, technical manufacture of types, 28–43
Nypro Ltd, 15

Organosols, use of nylons in, 203
Orientation, effects of, 102–104
Osmometry, 209–210
Oxidising agents, effect on nylons, 67

Oxime production for Beckmann rearrangement, 18

Packaging, nylon, 197
Perlon, 4
Perluranborsten, 3
Permeability to fluids, 62–64
Permittivity, 139–140
Phenol feedstock, 14–15
  for cyclohexanone, 16
Photosensitive resins, 204
Plasticisation, 68
PNC process, 19
Polyaddition, 32
Polyamides from aromatic diamines, 43–44
Polycondensation, 32
Polymerisation,
  above-the-melt, 176
  anionic, 33, 53, 173–176
  continuous (of 66), 30
  hydrolytic, 53
  interfacial, 51
  low-temperature, 50–52
  melt, 49
  ring-scission, 53
  solid-state, 49–50
  solution, 52
  step-reaction, 27, 52–53
Post-moulding treatment, 159–160
Powdered polyamides, uses of, 178–182
Pressing and sintering, 181–182
  particle size for, 181
  properties, comparison with injection moulding, 182

Raschig process for hydroxylamine, 17
Refractometry, differential, 212
Resistivity, 138–139
Rotating components, nylon, 193

Sachse-Mohr theory, 4
Salt, nylon 66, 12
Sandwich-moulding process, 9
SNIA Viscosa process, 18–19
Solid-phase forming, 9
Solvents for nylons, 68
Specific gravity, 132
Spherulites, 60–62
Stabilisers, action of, 77

Statistics, sales, production and usage, 4–7
Staudinger, H., 57
Stepwise polymerisation, 27, 52–53
Stereo-regularity, 60
Stress relaxation, 97
Super-polymers, 2
Swelling, 68

Techni-Chem process, 21
Telecommunications, use of nylon in, 195
Tensile properties, 79–80
Terminology, 10
Testing,
   accelerated, 75–77
   for design, 224
   for quality, 221–222
   for type, 222–223
Thermal properties, 132–137
   effect of fillers, 135
Threading and tapping, 185
Tracking resistance, 141
Transition points, 133
Tube extrusion, 168–170
Turning operations, 184

Type characteristics of nylons, 148–149

Ultra-violet spectra, use in analysis, 215–216

Vacuum extraction, 163
Vegetable oils, polymerised, nylons from, 42–43
Venting of moulds, 155
Viscosity,
   melt, 58, 208–209
   solution, 56–57, 206–208
   terminology, 57
VK process, 37–38

Wear of polyamides, 108–114
   effect of moisture, 129–130
   rates compared with metals, 109
   testing for, 110–114
Weather-O-Meter, 7
Wire-coating extrusion, 170–171

X-ray diffraction, 212, 213
Xenotest, 76